COMPLEX VARIABLES

A PHYSICAL APPROACH
WITH APPLICATIONS

SECOND EDITION

Textbooks in Mathematics

Series editors:
Al Boggess and Ken Rosen

COMPLEX VARIABLES
A PHYSICAL APPROACH
WITH APPLICATIONS

SECOND EDITION

STEVEN G. KRANTZ

CRC Press
Taylor & Francis Group
Boca Raton London New York

CRC Press is an imprint of the
Taylor & Francis Group, an **informa** business

Cover graphic created by Professor Elias Wegert of the Technische Universität Bergakademie Freiberg. The graphic is used with his permission.

CRC Press
Taylor & Francis Group
6000 Broken Sound Parkway NW, Suite 300
Boca Raton, FL 33487-2742

First issued in paperback 2022

© 2019 by Taylor & Francis Group, LLC
CRC Press is an imprint of Taylor & Francis Group, an Informa business

No claim to original U.S. Government works

Version Date: 20190325

ISBN 13: 978-1-03-247569-1 (pbk)
ISBN 13: 978-0-367-22267-3 (hbk)

DOI: 10.1201/9780429275166

Library of Congress Cataloging-in-Publication Data

Names: Krantz, Steven G. (Steven George), 1951- author.
Title: Complex variables : a physical approach with applications / Steven G. Krantz.
Description: Second edition. | Boca Raton : CRC Press, Taylor & Francis Group, 2019. | Includes bibliographical references and index.
Identifiers: LCCN 2018061709 | ISBN 9780367222673 (alk. paper)
Subjects: LCSH: Functions of complex variables. | Functions of complex variables--Textbooks. | Numbers, Complex--Textbooks. | Mathematical analysis--Textbooks.
Classification: LCC QA331.7 .K732 2019 | DDC 515/.9--dc23
LC record available at https://lccn.loc.gov/2018061709

Visit the Taylor & Francis Web site at
http://www.taylorandfrancis.com

and the CRC Press Web site at
http://www.crcpress.com

To the memory of Lars Ahlfors.

Contents

5 Applications of the Cauchy Theory   75

6 Isolated Singularities   95

Preface to the Second Edition for the Instructor

An earlier edition of this book has enjoyed notable success, and it is a pleasure now to prepare this new edition.

Complex analysis is still a central part of modern analytical thinking. It is used in engineering, physics, mathematics, astrophysics, and many other fields. It provides powerful tools for doing mathematical analysis, and often yields pleasing and unanticipated answers.

It is the purpose of this book to make the subject of complex analysis accessible to a broad audience of undergraduates. This will include engineering students, mathematics students, and many others. We present the key ideas of basic complex analysis without all the theoretical niceties that would be included in a graduate course. We do not prove the hard theorems; and, when we do prove a theorem, we do not use the word "proof." The idea is to make the subject user-friendly. A proof is a subjective device for convincing someone that something is true. Calling it a "proof" does not aid in the process, and runs the risk of alienating the audience.

This book has an exceptionally large number of examples and a large number of figures. Complex analysis, properly viewed, is a quite visual subject. There is an exercise set at the end of each section. Challenging problems are marked with a (∗). At the end of the book are solutions to selected exercises.

One of the main thrusts of the book is this: We present complex analysis as a natural outgrowth of the calculus (something that the students will already know quite well). It is not a new language, or a new way of thinking. Instead, it is an extension of material already mastered. This approach sets the text apart from others in the market, and should make the book more appealing to a larger group of students and instructors.

Of course an undergraduate course on complex analysis can and should contain plenty of applications. This book fills that need decisively. The applications come from physics, engineering, and other parts of science and technology. Included among these are

- the heat equation

- the wave equation

- the Dirichlet problem on the unit disc

- steady-state heat distribution on a lens-shaped region

- electrostatics on a disc

- incompressible fluid flow

- solving a differential equation with the Laplace transform

- population growth using the z-transform

- the Poisson integral formula

- several different proofs of the Fundamental Theorem of Algebra

This is only a sampling of the many topics that we cover in this text.

The complex numbers are a somewhat mysterious number system that seems to come out of the blue. It is important for students to see that this is really a very concrete set of objects that has very palpable and meaningful applications. We have meaningful applications throughout the book.

One unifying theme in this book is partial differential equations. The Cauchy–Riemann equations and the Laplacian have central roles in our exposition. Other differential equations also make an appearance.

The thrust of the exercises in the book, and of the text itself, is *not* to teach students the theory of complex analysis. Rather, it is to teach them how to handle complex numbers and complex functions with facility and grace. This is what people with a practical use for mathematics need to know.

I am pleased to thank the many fine reviewers who contributed their wisdom to help develop this book. I particularly thank Ken Rosen for his keen insights. I am grateful to my editor Robert Ross for his encouragement and support. I am also grateful to my readers, and look forward to hearing from them in the future.

Steven G. Krantz
St. Louis, Missouri

Preface to the Second Edition for the Student

Complex analysis may have always seemed rather mysterious to you. Where do the complex numbers come from? Why do they behave the way they do? Why do they endow negative real numbers with square roots?

If complex analysis is developed properly, then all these questions are answered in a comfortable and natural way. This book will do that job for you. It presents complex analysis as an outgrowth of the calculus that you have already mastered. Complex analysis is not some mysterious new world. It is simply an extension of things that you already know.

An important aspect of this book is that it develops a great many applications of complex analysis. These applications come from physics, engineering, and other parts of science and technology. Included among these are

- the heat equation

- the wave equation

- the Dirichlet problem on the unit disc

- steady-state heat distribution on a lens-shaped region

- electrostatics on a disc

- incompressible fluid flow

- solving a differential equation with the Laplace transform

- population growth using the z-transform

- the Poisson integral formula

This is only a sampling of the many topics that we cover in this text.

The book has many examples and graphics in order to give you a tactile feel for the subject matter. There are exercises at all levels—from drill exercises to thought exercises to speculative exercises. Naturally solutions to selected exercises appear at the end of the book.

In order to make the book most useful for you we have included a Glossary and a Table of Notation. There are also a Table of Laplace Transforms and a Guide to the Literature included. The first of these will help you in your work in the book. The second will be useful as you become more and more fascinated with complex variables and want to continue your reading.

In sum, this is a comprehensive introduction to the field of complex variables—a field that you need to know and understand in order to be able to develop in the mathematical sciences, in engineering, and in technology.

We wish you good fortune in your study of complex analysis and your education as a mathematical scientist.

Steven G. Krantz
St. Louis, Missouri

Preface to the First Edition

Complex variables is one of the grand old ladies of mathematics. Originally conceived in the pursuit of solutions of polynomial equations, complex variables blossomed in the hands of Euler, Argand, and others into the free-standing subject of complex analysis.

Like the negative numbers and zero, complex numbers were at first viewed with some suspicion. To be sure, they were useful tools for solving certain types of problems. But what were they precisely and where did they come from? What did they correspond to in the real world?

Today we have a much more concrete, and more catholic, view of the matter. First of all, we now know how to construct the complex numbers using rigorous mathematical techniques. Second, we understand how complex eigenvalues arise in the study of mechanical vibrations, how complex functions model incompressible fluid flow, and how complex variables enable the Laplace transform and the solution of a variety of differential equations that arise from physics and engineering.

It is essential for the modern undergraduate engineering student, as well as the math major and the physics major, to understand the basics of complex variable theory. The need then is for a textbook that presents the elements of the subject while requiring only a solid background in the calculus of one and several variables. This is such a text. Of course there are other solid books for such a course. The book of Brown and Churchill has stood for many editions. The book of Saff and Snider, a more recent offering, is well-written and incisive. What makes the present text distinctive are the following features:

(1) We work in ideas from physics and engineering beginning in Chapter 1, and continuing throughout the book. Applications are an integral part of the presentation at every stage.

(2) Every chapter contains exercises that illustrate the applications.

(3) A very important attribute (and one not represented in any other book, as few other authors are qualified to make such a presentation) is that this text presents the subject of complex analysis as a natural continuation of the calculus. Most complex analysis texts exhibit the subject as a free-standing collection of ideas, independent of other parts of mathematical analysis and having its own body of techniques and tricks. This is in fact a misrepresentation of the subject and leads to copious misunderstanding and misuse of the ideas. We are able to present complex analysis as part and parcel of the world view that the student has developed in his/her earlier course work. The result is that students can master the material more effectively and use it with good result in other courses in engineering and physics.

(4) The book will have stimulating exercises at the three levels of drill, exploration, and theory. There will be a comfortable balance between theory and applications.

(5) Every section will have examples that illustrate both the theory and the practice of complex variables.

(6) The book will have copious figures.

(7) We will use differential equations as a unifying theme throughout the book.

The subject of complex variables has many aspects—from the algebraic features of a complete number field, to the analytic properties imposed by the Cauchy integral formula, to the geometric qualities coming from the idea of conformality. The student must be acqainted with all components of the field. This text speaks all the languages, and shows the student how to deal with all the different approaches to complex analysis. The examples illustrate all the key concepts; the exercises reinforce the basic skills, and give practice in all the fundamental ideas.

Complex variables is a living, breathing part of modern mathematics. It is a vibrant area of mathematical research in its own right, and it interacts fruitfully with harmonic analysis, partial differential equations, Fourier series, group representations, and many other parts of the mathematical sciences. And it is an essential tool in applications. This text will illustrate and teach all facets of the subject in a lively manner that will speak to the needs of modern students. It will give them a powerful toolkit for future work in the mathematical sciences, and will also point to new directions for additional learning.

— SGK
St. Louis, Missouri

Chapter 1

Basic Ideas

1.1 Complex Arithmetic

1.1.1 The Real Numbers

The real number system consists of both the rational numbers (numbers with terminating or repeating decimal expansions) and the irrational numbers (numbers with infinite, non-repeating decimal expansions). The real numbers are denoted by the symbol $\mathbb{R}$. We let $\mathbb{R}^2 = \{(x, y) : x \in \mathbb{R}, \ y \in \mathbb{R}\}$ (Figure 1.1).

1.1.2 The Complex Numbers

The complex numbers $\mathbb{C}$ consist of $\mathbb{R}^2$ equipped with some special algebraic operations. One defines

$$
\begin{aligned}
(x, y) + (x', y') &= (x + x', y + y'), \\
(x, y) \cdot (x', y') &= (xx' - yy', xy' + yx').
\end{aligned}
$$

These operations of $+$ and $\cdot$ are commutative and associative.

EXAMPLE 1 We may calculate that

$$(3, 7) + (2, -4) = (3 + 2, 7 + (-4)) = (5, 3).$$

Also

$$(3, 7) \cdot (2, -4) = (3 \cdot 2 - 7 \cdot (-4), 3 \cdot (-4) + 7 \cdot 2) = (34, 2).$$

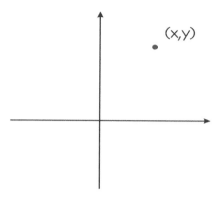

Figure 1.1: A point in the plane.

□

Of course we sometimes wish to subtract complex numbers. We define

$$z - w = z + (-w).$$

Thus if $z = (11, -6)$ and $w = (1, 4)$ then

$$z - w = z + (-w) = (11, -6) + (-1, -4) = (10, -10).$$

We denote $(1, 0)$ by 1 and $(0,1)$ by i. We also denote $(0, 0)$ by 0. If $\alpha \in \mathbb{R}$, then we identify α with the complex number $(\alpha, 0)$. Using this notation, we see that

$$\alpha \cdot (x, y) = (\alpha, 0) \cdot (x, y) = (\alpha x, \alpha y). \tag{1.1}$$

In particular,

$$1 \cdot (x, y) = (1, 0) \cdot (x, y) = (x, y).$$

We may calculate that

$$x \cdot 1 + y \cdot i = (x, 0) \cdot (1, 0) + (y, 0) \cdot (0, 1) = (x, 0) + (0, y) = (x, y). \tag{1.2}$$

Thus every complex number (x, y) can be written in one and only one fashion in the form $x \cdot 1 + y \cdot i$ with $x, y \in \mathbb{R}$. We usually write the number even more succinctly as $x + iy$.

EXAMPLE 2 The complex number $(-2, 5)$ is usually written as

$$(-2, 5) = -2 + 5i.$$

The complex number $(4, 9)$ is usually written as

$$(4, 9) = 4 + 9i.$$

The complex number $(-3, 0)$ is usually written as

$$(-3, 0) = -3 + 0i = -3.$$

The complex number $(0, 6)$ is usually written as

$$(0, 6) = 0 + 6i = 6i.$$

□

In this more commonly used notation, the laws of addition and multiplication become

$$
\begin{aligned}
(x + iy) + (x' + iy') &= (x + x') + i(y + y'), \\
(x + iy) \cdot (x' + iy') &= (xx' - yy') + i(xy' + yx').
\end{aligned}
$$

Observe that $i \cdot i = -1$. Indeed,

$$i \cdot i = (0, 1) \cdot (0, 1) = (0 \cdot 0 - 1 \cdot 1) + i(0 \cdot 1 + 1 \cdot 0) = -1 + 0i = -1.$$

This is historically the most important fact about the complex numbers—that they provide negative numbers with square roots. More generally, the complex numbers provide *any polynomial equation* with roots. We shall develop these ideas in detail below.

Certainly our multiplication law is consistent with the scalar multiplication introduced in line (1.1).

Insight: The multiplicative law presented at the beginning of §§1.1.2 may at first seem strange and counter-intuitive. Why not take the simplest possible route and define

$$(x, y) \cdot (x', y') = (xx', yy') \, ? \tag{1.3}$$

This would certainly be easier to remember, and is consistent with what one might guess. The trouble is that definition (1.3), while simple, has a number of liabilities. First of all, it would lead to

$$(1,0) \cdot (0,1) = (0,0) = 0 \,.$$

Thus we would have the product of two nonzero numbers equaling zero—an eventuality that we want to always avoid in any arithmetic. Second, the main point of the complex numbers is that we want a negative number to have a square root. That would not happen if (1.3) were our definition of multiplication.

The definition at the start of §§1.1.2 is in fact a very clever idea that creates a *new number system* with many marvelous new properties. The purpose of this text is to acquaint you with this new world. □

EXAMPLE 3 The fact that $i \cdot i = -1$ means that the number -1 has a square root. This fact is at first counterintuitive. If we stick to the real number system, then only nonnegative numbers have square roots. In the complex number system, *any* number has a square root—in fact two of them.[1] For example,

$$(1+i)^2 = 2i$$

and

$$(-1-i)^2 = 2i \,.$$

Later in this chapter we will learn how to find both the square roots, and in fact all n of the nth roots, of any complex number. □

The symbols z, w, ζ are frequently used to denote complex numbers. We usually take $z = x + iy$, $w = u + iv$, $\zeta = \xi + i\eta$. The real number x is called the *real part* of z and is written $x = \operatorname{Re} z$. The real number y is called the *imaginary part* of z and is written $y = \operatorname{Im} z$.

---

[1]The number 0 has just one square root. It is the only root of the polynomial equation $z^2 = 0$. All other complex numbers α have two distinct square roots. They are the roots of the polynomial equation $z^2 = \alpha$ or $z^2 - \alpha = 0$. The matter will be treated in greater detail below. In particular, we shall be able to put these ideas in the context of the fundamental theorem of algebra.

EXAMPLE 4 The real part of the complex number $z = 4 - 8i$ is 4. We write

$$\operatorname{Re} z = 4\,.$$

The imaginary part of z is -8. We write

$$\operatorname{Im} z = -8\,.$$

□

EXAMPLE 5 Addition of complex numbers corresponds exactly to addition of vectors in the plane. Specifically, if $z = x + iy$ and $w = u + iv$ then

$$z + w = (x + u) + i(y + v)\,.$$

If we make the correspondence

$$z = x + iy \leftrightarrow \mathbf{z} = \langle x, y \rangle$$

and

$$w = u + iv \leftrightarrow \mathbf{w} = \langle u, v \rangle$$

then we have

$$\mathbf{z} + \mathbf{w} = \langle x, y \rangle + \langle u, v \rangle = \langle x + u, y + v \rangle\,.$$

Clearly

$$(x + u) + i(y + v) \leftrightarrow \langle x + u, y + v \rangle\,.$$

But complex multiplication *does not* correspond to any standard vector operation. Indeed it cannot. For the standard vector dot product has no concept of multiplicative inverse; and the standard vector cross product has no concept of multiplicative inverse. But one of the main points of the complex number operations is that they turn this number system into a *field*: every nonzero number does indeed have a multiplicative inverse. This is a very special property of two-dimensional space. There is no other Eucliean space (except of course the real line) that can be equipped with commutative operations of addition and multiplication so that **(i)** every number has an additive inverse and **(ii)** every nonzero number has a multiplicative inverse. We shall learn more about these ideas below. □

The complex number $x - iy$ is by definition the complex *conjugate* of the complex number $x + iy$. If $z = x + iy$, then we denote the conjugate[2] of z with the symbol $\bar{z}$; thus $\bar{z} = x - iy$.

1.1.3   Complex Conjugate

Note that $z + \bar{z} = 2x$, $z - \bar{z} = 2iy$. Also

$$\overline{z + w} = \bar{z} + \bar{w},$$

$$\overline{z \cdot w} = \bar{z} \cdot \bar{w}.$$

A complex number is real (has no imaginary part) if and only if $z = \bar{z}$. It is imaginary (has no real part) if and only if $z = -\bar{z}$.

EXAMPLE 6 Let $z = -7 + 6i$ and $w = 4 - 9i$. Then

$$\bar{z} = -7 - 6i$$

and

$$\bar{w} = 4 + 9i.$$

Notice that

$$\bar{z} + \bar{w} = (-7 - 6i) + (4 + 9i) = -3 + 3i,$$

and that number is exactly the conjugate of

$$z + w = -3 - 3i.$$

Notice also that

$$\bar{z} \cdot \bar{w} = (-7 - 6i) \cdot (4 + 9i) = 26 - 87i,$$

and that number is exactly the conjugate of

$$z \cdot w = 26 + 87i.$$

$\square$

---

[2]Rewriting history a bit, we may account for the concept of "conjugate" as follows. If $p(z) = az^2 + bz + c$ is a polynomial with real coefficients, and if $z = x + iy$ is a root of this polynomial, then $\bar{z} = x - iy$ will also be a root of that same polynomial. This assertion is immediate from the quadratic formula, or by direct calculation. Thus $x + iy$ and $x - iy$ are *conjugate roots* of the polynomial p.

Exercises

1.   Let $z = 13 + 5i$, $w = 2 - 6i$, and $\zeta = 1 + 9i$. Calculate $z + w$, $w - \zeta$, $z \cdot \zeta$, $w \cdot \zeta$, and $\zeta - z$.

2.   Let $z = 4 - 7i$, $w = 1 + 3i$, and $\zeta = 2 + 2i$. Calculate $\bar{z}$, $\bar{\zeta}$, $\overline{z - w}$, $\overline{\zeta + z}$, $\overline{\zeta \cdot w}$.

3.   If $z = 6 - 2i$, $w = 4 + 3i$, and $\zeta = -5 + i$, then calculate $z + \bar{z}$, $z + 2\bar{z}$, $z - \bar{w}$, $z \cdot \bar{\zeta}$, and $w \cdot \bar{\zeta}^2$.

4.   If z is a complex number then $\bar{z}$ has the same distance from the origin as z. Explain why.

5.   If z is a complex number then $\bar{z}$ and z are situated symmetrically with respect to the x-axis. Explain why.

6.   If z is a complex number then $-\bar{z}$ and z are situated symmetrically with respect to the y-axis. Explain why.

7.   Explain why addition in the real numbers is a special case of addition in the complex numbers. Explain why the two operations are logically consistent.

8.   Explain why multiplication in the real numbers is a special case of multiplication in the complex numbers. Explain why the two operations are logically consistent.

1.2   Algebraic and Geometric Properties

1.2.1   Modulus of a Complex Number

The ordinary Euclidean distance of (x, y) to $(0, 0)$ is $\sqrt{x^2 + y^2}$ (Figure 1.2). We also call this number the *modulus* of the complex number $z = x + iy$ and we write $|z| = \sqrt{x^2 + y^2}$. Note that

$$z \cdot \bar{z} = x^2 + y^2 = |z|^2. \tag{1.4}$$

The distance from z to w is $|z - w|$. We also have the easily verified formulas $|zw| = |z||w|$ and $|\mathrm{Re}\, z| \leq |z|$ and $|\mathrm{Im}\, z| \leq |z|$.

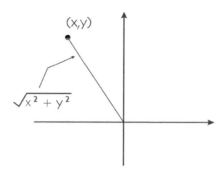

Figure 1.2: Distance to the origin or modulus.

The very important *triangle inequality* says that

$$|z + w| \leq |z| + |w|.$$

We shall discuss this relation in greater detail below. For now, the interested reader may wish to square both sides, cancel terms, and see what the inequality reduces to.

EXAMPLE 7 The complex number $z = 7 - 4i$ has modulus given by

$$|z| = \sqrt{7^2 + (-4)^2} = \sqrt{65}.$$

The complex number $w = 2 + i$ has modulus given by

$$|w| = \sqrt{2^2 + 1^2} = \sqrt{5}.$$

Finally, the complex number $z + w = 9 - 3i$ has modulus given by

$$|z + w| = \sqrt{9^2 + (-3)^2} = \sqrt{90}.$$

According to the triangle inequality,

$$|z + w| \leq |z| + |w|,$$

and we may now confirm this arithmetically as

$$\sqrt{90} \leq \sqrt{65} + \sqrt{5}.$$

$\square$

1.2.2 The Topology of the Complex Plane

If P is a complex number and $r > 0$, then we set

$$D(P, r) = \{z \in \mathbb{C} : |z - P| < r\}$$

and

$$\overline{D}(P, r) = \{z \in \mathbb{C} : |z - P| \leq r\}.$$

The first of these is the *open disc with center P and radius r*; the second is the *closed disc with center P and radius r* (Figure 1.3). We often use the simpler symbols D and $\overline{D}$ to denote, respectively, the discs $D(0, 1)$ and $\overline{D}(0, 1)$.

We say that a set $U \subseteq \mathbb{C}$ is *open* if, for each $P \in U$, there is an $r > 0$ such that $D(P, r) \subseteq U$. Thus an open set is one with the property that each point P of the set is surrounded by neighboring points (that is, the points of distance less than r from P) that are still in the set—see Figure 1.4. Of course the number r will depend on P. As examples, $U = \{z \in \mathbb{C} : \mathrm{Re}\, z > 1\}$ is open, but $F = \{z \in \mathbb{C} : \mathrm{Re}\, z \leq 1\}$ is not (Figure 1.5). Observe that, in these figures, we use a *solid line* to indicate that the boundary is included in the set; we use a *dotted line* to indicate that the boundary is not included in the set. A figure like Figure 1.5 that depicts the complex numbers in an x-y plane is called an *Argand diagram*.

A set $E \subseteq \mathbb{C}$ is said to be *closed* if $\mathbb{C} \setminus E \equiv \{z \in \mathbb{C} : z \notin E\}$ (the complement of E in $\mathbb{C}$) is open. The set F in the last paragraph is closed.

It is *not* the case that any given set is either open or closed. For example, the set $W = \{z \in \mathbb{C} : 1 < \mathrm{Re}\, z \leq 2\}$ is *neither* open *nor* closed (Figure 1.6).

We say that a set $E \subset \mathbb{C}$ is *connected* if there do not exist non-empty disjoint open sets U and V such that $E = (U \cap E) \cup (V \cap E)$. Refer to Figure 1.7 for these ideas. It is a useful fact that if E is an open set, then E is connected if and only if it is path-connected; this means that any two points of E can be connected by a continuous path or curve that lies entirely in the set. See Figure 1.8.

1.2.3 The Complex Numbers as a Field

Let 0 denote the complex number $0 + i0$. If $z \in \mathbb{C}$, then $z + 0 = z$. Also, letting $-z = -x - iy$, we have $z + (-z) = 0$. So every complex number

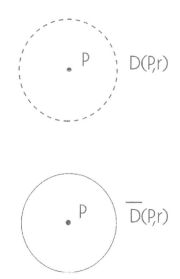

Figure 1.3: An open disc and a closed disc.

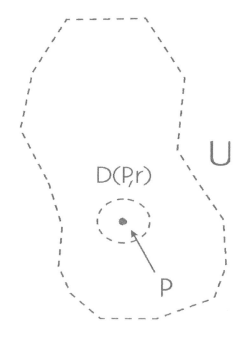

Figure 1.4: An open set.

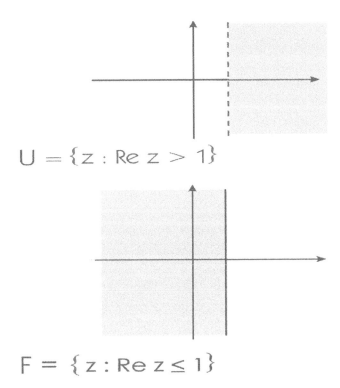

Figure 1.5: An open set and a non-open set.

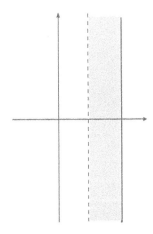

Figure 1.6: A set that is neither open nor closed.

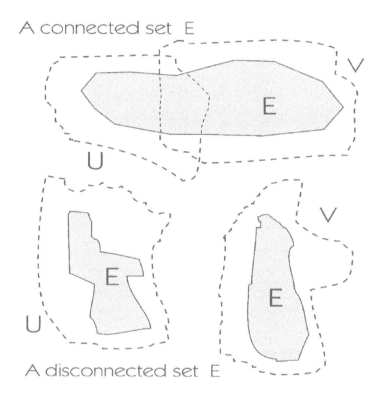

Figure 1.7: A connected set and a disconnected set.

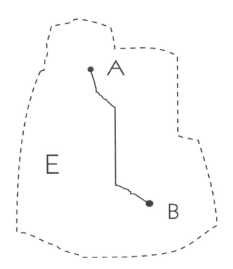

Figure 1.8: An open set is connected if and only if it is path-connected.

has an additive inverse, and that inverse is unique. One may also readily verify that $0 \cdot z = z \cdot 0 = 0$ for any complex number z.

Since $1 = 1 + i0$, it follows that $1 \cdot z = z \cdot 1 = z$ for every complex number z. If $z \neq 0$, then $|z|^2 \neq 0$ and

$$z \cdot \left(\frac{\overline{z}}{|z|^2} \right) = \frac{|z|^2}{|z|^2} = 1. \tag{1.5}$$

So every non-zero complex number has a multiplicative inverse, and that inverse is unique. It is natural to define $1/z$ to be the multiplicative inverse $\overline{z}/|z|^2$ of z and, more generally, to define

$$\frac{z}{w} = z \cdot \frac{1}{w} = \frac{z\overline{w}}{|w|^2} \qquad \text{for } w \neq 0. \tag{1.6}$$

We also have $\overline{z/w} = \overline{z}/\overline{w}$.

EXAMPLE 8 The idea of multiplicative inverse is at first counterintuitive. So let us look at a specific instance.

Let $z = 2 + 3i$. It is all too easy to say that the multiplicative inverse of z is

$$\frac{1}{z} = \frac{1}{2 + 3i}.$$

The trouble is that, as written, $1/(2 + 3i)$ is *not* a complex number. Recall that a complex number is a number of the form $x + iy$. But our discussion preceding this example enables us to clarify the matter.

Because in fact the multiplicative inverse of $2 + 3i$ is

$$\frac{\overline{z}}{|z|^2} = \frac{2 - 3i}{13} \, .$$

The advantage of looking at things this way is that the multiplicative inverse is in fact now a complex number; it is

$$\frac{2}{13} - i\frac{3}{13} \, .$$

And we may check directly that this number does the job:

$$(2 + 3i) \cdot \left(\frac{2}{13} - i\frac{3}{13} \right) = \left(2 \cdot \frac{2}{13} + 3 \cdot \frac{3}{13} \right) + i \left(2 \cdot \left(-\frac{3}{13} \right) + 3 \cdot \frac{2}{13} \right)$$
$$= 1 + 0i$$
$$= 1 \, . \qquad \qquad \square$$

Multiplication and addition satisfy the usual distributive, associative, and commutative laws. Therefore $\mathbb{C}$ is a *field* (see [HER]). The field $\mathbb{C}$ contains a copy of the real numbers in an obvious way:

$$\mathbb{R} \ni x \mapsto x + i0 \in \mathbb{C}. \tag{1.7}$$

This identification respects addition and multiplication. So we can think of $\mathbb{C}$ as a field extension of $\mathbb{R}$: it is a larger field which contains the field $\mathbb{R}$.

1.2.4   The Fundamental Theorem of Algebra

It is not true that every non-constant polynomial with real coefficients has a real root. For instance, $p(x) = x^2 + 1$ has no real roots. The Fundamental Theorem of Algebra states that every polynomial with complex coefficients has a complex root (see the treatment in §§1.2.4, 5.1.4, 9.3.3 below). The complex field $\mathbb{C}$ is the *smallest* field that contains $\mathbb{R}$ and has this so-called algebraic closure property.

Exercises

1. Let $z = 6 - 9i$, $w = 4 + 2i$, $\zeta = 1 + 10i$. Calculate $|z|$, $|w|$, $|z + w|$, $|\zeta - w|$, $|z \cdot w|$, $|z + w|$, $|\zeta \cdot z|$. Confirm directly that

$$|z + w| \leq |z| + |w|,$$

$$|z \cdot w| = |z||w|,$$

$$|\zeta \cdot z| = |\zeta||z|.$$

2. Find complex numbers z, w such that $|z| = 5$, $|w| = 7$, $|z + w| = 9$.

3. Find complex numbers z, w such that $|z| = 1$, $|w| = 1$, and $z/w = i^3$.

4. Let $z = 4 - 6i$, $w = 2 + 7i$. Calculate z/w, w/z, and $1/w$.

5. Sketch these discs on the same set of axes: $D(2 + 3i, 4)$, $D(1 - 2i, 2)$, $\overline{D}(i, 5)$, $\overline{D}(6 - 2i, 5)$.

6. Which of these sets is open? Which is closed? Why or why not?

   (a) $\{x + iy \in \mathbb{C} : x^2 + 4y^2 \leq 4\}$
   (b) $\{x + iy \in \mathbb{C} : x < y\}$
   (c) $\{x + iy \in \mathbb{C} : 2 \leq x + y < 5\}$
   (d) $\{x + iy \in \mathbb{C} : 4 < \sqrt{x^2 + 3y^2}\}$
   (e) $\{x + iy \in \mathbb{C} : 5 \leq \sqrt{x^4 + 2y^6}\}$

7. Consider the polynomial $p(z) = z^3 - z^2 + 2z - 2$. How many real roots does p have? How many complex roots? Explain.

8. The polynomial $q(z) = z^3 - 3z + 2$ is of degree three, yet it does *not* have three distinct roots. Explain.

9. Write a fourth degree polynomial $q(z)$ whose roots are 1, -1, i, and $-i$. These four numbers are all the fourth roots of 1. Explain therefore why q has such a simple form.

Chapter 2

The Exponential and Applications

2.1   The Exponential Function

We define the complex exponential as follows:

(2.1) If $z = x$ is real, then

$$e^z = e^x \equiv \sum_{j=0}^{\infty} \frac{x^j}{j!}$$

as in calculus. Here ! denotes the usual "factorial" operation:

$$j! = j \cdot (j-1) \cdot (j-2) \cdots 3 \cdot 2 \cdot 1 \,.$$

By custom, $0! = 1$.

(2.2) If $z = iy$ is pure imaginary, then

$$e^z = e^{iy} \equiv \cos y + i \sin y.$$

(2.3) If $z = x + iy$, then

$$e^z = e^{x+iy} \equiv e^x \cdot e^{iy} = e^x \cdot (\cos y + i \sin y).$$

This tri-part definition may seem a bit mysterious. But we may justify it formally as follows (a detailed discussion of complex power series will come later). Consider the definition

$$e^z = \sum_{j=0}^{\infty} \frac{z^j}{j!} . \tag{2.4}$$

This is a natural generalization of the familiar definition of the exponential function from calculus.

We may write this out as

$$e^z = 1 + z + \frac{z^2}{2!} + \frac{z^3}{3!} + \frac{z^4}{4!} + \cdots . \tag{2.5}$$

In case $z = x$ is real, this gives the familiar

$$e^x = 1 + x + \frac{x^2}{2!} + \frac{x^3}{3!} + \frac{x^4}{4!} + \cdots . $$

In case $z = iy$ is pure imaginary, then (2.5) gives

$$e^{iy} = 1 + iy - \frac{y^2}{2!} - i\frac{y^3}{3!} + \frac{y^4}{4!} + i\frac{y^5}{5!} - \frac{y^6}{6!} - i\frac{y^7}{7!} + - \cdots . \tag{2.6}$$

Grouping the real terms and the imaginary terms we find that

$$e^{iy} = \left[1 - \frac{y^2}{2!} + \frac{y^4}{4!} - \frac{y^6}{6!} + - \cdots \right] + i\left[y - \frac{y^3}{3!} + \frac{y^5}{5!} - \frac{y^7}{7!} + - \cdots \right] = \cos y + i \sin y . \tag{2.7}$$

This is the same as the definition that we gave above in **(1.8)**.

Part **(2.3)** of the definition is of course justified by the usual rules of exponentiation.

An immediate consequence of this new definition of the complex exponential is the following complex-analytic definition of the sine and cosine functions:

$$\cos z = \frac{e^{iz} + e^{-iz}}{2} , \tag{2.8}$$

$$\sin z = \frac{e^{iz} - e^{-iz}}{2i} . \tag{2.9}$$

Note that when $z = x + i0$ is real this new definition is consistent with the Euler formula:

$$e^{ix} = \cos x + i \sin x. \tag{2.10}$$

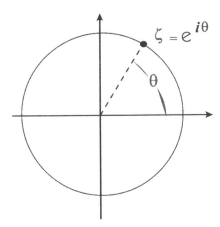

Figure 2.1: Polar coordinates of a point in the plane.

2.1.1 Laws of Exponentiation

The complex exponential satisfies familiar rules of exponentiation:

$$e^{z+w} = e^z \cdot e^w \quad \text{and} \quad (e^z)^w = e^{zw}. \tag{2.11}$$

Also

$$(e^z)^n = \underbrace{e^z \cdots e^z}_{n \text{ times}} = e^{nz}. \tag{2.12}$$

2.1.2 The Polar Form of a Complex Number

A consequence of our first definition of the complex exponential—see **(2.2)**—is that if $\zeta \in \mathbb{C}$, $|\zeta| = 1$, then there is a unique number θ, $0 \leq \theta < 2\pi$, such that $\zeta = e^{i\theta}$ (see Figure 2.1). Here θ is the (signed) angle between the positive x axis and the ray $\overrightarrow{0\zeta}$.

Now if z is any non-zero complex number, then

$$z = |z| \cdot \left(\frac{z}{|z|} \right) \equiv |z| \cdot \zeta \tag{2.13}$$

where $\zeta \equiv z/|z|$ has modulus 1. Again, letting θ be the angle between the positive real axis and $\overrightarrow{0\zeta}$, we see that

$$
\begin{aligned}
z &= |z| \cdot \zeta \\
&= |z| \cdot e^{i\theta} \\
&= re^{i\theta} \; ,
\end{aligned}
\tag{2.14}
$$

where $r = |z|$. This form is called the *polar* representation for the complex number z. (Note that some classical books write the expression $z = re^{i\theta} = r(\cos\theta + i\sin\theta)$ as $z = r\mathrm{cis}\,\theta$. The reader should be aware of this notation, though we shall not use it in this book.)

EXAMPLE 1 Let $z = 1 + \sqrt{3}i$. Then $|z| = \sqrt{1^2 + (\sqrt{3})^2} = 2$. Hence

$$
z = 2 \cdot \left(\frac{1}{2} + i\,\frac{\sqrt{3}}{2} \right).
\tag{2.15}
$$

The number in parentheses is of unit modulus and subtends an angle of $\pi/3$ with the positive x-axis. Therefore

$$
1 + \sqrt{3}i = z = 2 \cdot e^{i\pi/3}.
\qquad \square
$$

It is often convenient to allow angles that are greater than or equal to 2π in the polar representation; when we do so, the polar representation is no longer unique. For if k is an integer, then

$$
\begin{aligned}
e^{i\theta} &= \cos\theta + i\sin\theta \\
&= \cos(\theta + 2k\pi) + i\sin(\theta + 2k\pi) \\
&= e^{i(\theta + 2k\pi)} \; .
\end{aligned}
\tag{2.16}
$$

Exercises

1. Calculate (with your answer in the form $a + ib$) the values of $e^{\pi i}$, $e^{(\pi/3)i}$, $5e^{-i(\pi/4)}$, $2e^{i}$, $7e^{-3i}$.

2. Write these complex numbers in polar form: $2 + 2i$, $1 + \sqrt{3}i$, $\sqrt{3} - i$, $\sqrt{2} - i\sqrt{2}$, i, $-1 - i$.

3. If $e^z = 2 - 2i$ then what can you say about z? (**Hint:** There is more than one answer.)

4. If $w^5 = z$ and $|z| = 3$ then what can you say about $|w|$?

5. If $w^5 = z$ and z subtends an angle of $\pi/4$ with the positive x-axis, then what can you say about the angle that w subtends with the positive x-axis? [**Hint:** There is more than one answer to this question.]

6. If $w^2 = z^3$ then how are the polar forms of z and w related?

7. Write all the polar forms of the complex number $-\sqrt{2} + i\sqrt{6}$.

8. If $z = re^{i\theta}$ and $w = se^{i\psi}$ then what can you say about the polar form of $z + w$? What about $z \cdot w$?

2.1.3 Roots of Complex Numbers

The properties of the exponential operation can be used, together with the polar representation, to find the nth roots of a complex number.

EXAMPLE 2 To find all sixth roots of 2, we let $re^{i\theta}$ be an arbitrary sixth root of 2 and solve for r and θ. If

$$\left(re^{i\theta}\right)^6 = 2 = 2 \cdot e^{i0} \tag{2.17}$$

or

$$r^6 e^{i6\theta} = 2 \cdot e^{i0}, \tag{2.18}$$

then it follows that $r = 2^{1/6} \in \mathbb{R}$ and $\theta = 0$ solve this equation. So the real number $2^{1/6} \cdot e^{i0} = 2^{1/6}$ is a sixth root of two. This is not terribly surprising, but we are not finished.

We may also solve

$$r^6 e^{i6\theta} = 2 = 2 \cdot e^{2\pi i}. \tag{2.19}$$

Notice that we are taking advantage of the ambiguity built into the polar representation: The number 2 may be written as $2 \cdot e^{i0}$, but it may also be written as $2 \cdot e^{2\pi i}$ or as $2 \cdot e^{4\pi i}$, and so forth.

Hence
$$r = 2^{1/6}\ ,\ \theta = 2\pi/6 = \pi/3. \tag{2.20}$$

This gives us the number

$$2^{1/6}e^{i\pi/3} = 2^{1/6}\left(\cos\pi/3 + i\sin\pi/3\right) = 2^{1/6}\left(\frac{1}{2} + i\frac{\sqrt{3}}{2}\right) \tag{2.21}$$

as a sixth root of two. Similarly, we can solve

$$\begin{aligned}
r^6 e^{i6\theta} &= 2 \cdot e^{4\pi i}\\
r^6 e^{i6\theta} &= 2 \cdot e^{6\pi i}\\
r^6 e^{i6\theta} &= 2 \cdot e^{8\pi i}\\
r^6 e^{i6\theta} &= 2 \cdot e^{10\pi i}
\end{aligned}$$

to obtain the other four sixth roots of 2:

$$2^{1/6}\left(-\frac{1}{2} + i\frac{\sqrt{3}}{2}\right) \tag{2.22}$$

$$-2^{1/6} \tag{2.23}$$

$$2^{1/6}\left(-\frac{1}{2} - i\frac{\sqrt{3}}{2}\right) \tag{2.24}$$

$$2^{1/6}\left(\frac{1}{2} - i\frac{\sqrt{3}}{2}\right). \tag{2.25}$$

These are in fact all the sixth roots of 2, because the sixth roots are the roots of the sixth degree polynomial $z^6 - 2$. □

EXAMPLE 3 Let us find all third roots of i. We begin by writing i as

$$i = e^{i\pi/2}. \tag{2.26}$$

Solving the equation
$$(re^{i\theta})^3 = i = e^{i\pi/2} \tag{2.27}$$

then yields $r = 1$ and $\theta = \pi/6$.

Next, we write $i = e^{i(\pi/2 + 2\pi)} = e^{i5\pi/2}$ and solve

$$(re^{i\theta})^3 = e^{i5\pi/2} \tag{2.28}$$

to obtain that $r = 1$ and $\theta = 5\pi/6$.

Lastly, we write $i = e^{i(\pi/2 + 4\pi)} = e^{i9\pi/2}$ and solve

$$(re^{i\theta})^3 = e^{i9\pi/2} \tag{2.29}$$

to obtain that $r = 1$ and $\theta = 9\pi/6 = 3\pi/2$.

In summary, the three cube roots of i are

$$e^{i\pi/6} = \frac{\sqrt{3}}{2} + i\frac{1}{2},$$

$$e^{i5\pi/6} = -\frac{\sqrt{3}}{2} + i\frac{1}{2},$$

$$e^{i3\pi/2} = -i. \qquad \square$$

It is worth it for you to take the time to sketch the six 6th roots of 2 (from Example 13) on a single set of axes. Also sketch all the third roots of i on a single set of axes. Observe that the six 6th roots of 2 are equally spaced about a circle that is centered at the origin and has radius $2^{1/6}$. Likewise, the three cube roots of i are equally spaced about a circle that is centered at the origin and has radius 1.

2.1.4 The Argument of a Complex Number

The (non-unique) angle θ associated to a complex number $z \neq 0$ is called its *argument*, and is written $\arg z$. For instance, $\arg(1 + i) = \pi/4$. See Figure 2.2. But it is also correct to write $\arg(1+i) = 9\pi/4, 17\pi/4, -7\pi/4$, etc. We generally choose the argument θ to satisfy $0 \leq \theta < 2\pi$. This is the *principal branch* of the argument.

Under multiplication of complex numbers, arguments are additive and moduli multiply. That is, if $z = re^{i\theta}$ and $w = se^{i\psi}$, then

$$z \cdot w = re^{i\theta} \cdot se^{i\psi} = (rs) \cdot e^{i(\theta+\psi)}. \tag{2.30}$$

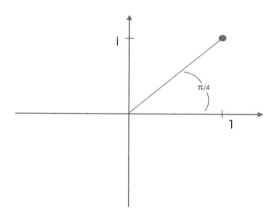

Figure 2.2: Polar form of the complex number $1 + i$.

2.1.5   Fundamental Inequalities

We next record a few inequalities.

The Triangle Inequality: If $z, w \in \mathbb{C}$, then

$$|z + w| \leq |z| + |w|. \tag{2.31}$$

More generally,

$$\left| \sum_{j=1}^{n} z_j \right| \leq \sum_{j=1}^{n} |z_j|. \tag{2.32}$$

For the verification, square both sides. We obtain

$$|z + w|^2 \leq (|z| + |w|)^2$$

or

$$(z + w) \cdot (\overline{z + w}) \leq (|z| + |w|)^2.$$

Multiplying this out yields

$$|z|^2 + z\overline{w} + w\overline{z} + |w|^2 \leq |z|^2 + 2|z||w| + |w|^2.$$

Cancelling like terms yields

$$2\mathrm{Re}\,(z\overline{w}) \leq 2|z||w|$$

or
$$\mathrm{Re}\,(z\overline{w}) \leq |z||w|\,.$$

It is convenient to rewrite this as
$$\mathrm{Re}\,(z\overline{w}) \leq |z\overline{w}|\,. \tag{2.33}$$

But it is true, for any complex number ζ, that $|\mathrm{Re}\,\zeta| \leq |\zeta|$. Our argument runs both forward and backward. So (2.33) implies (2.31). This establishes the basic triangle inequality.

To give an idea of why the more general triangle inequality is true, consider just three terms. We have
$$\begin{aligned}
|z_1 + z_2 + z_3| &= |z_1 + (z_2 + z_3)|\\
&\leq |z_1| + |z_2 + z_3|\\
&\leq |z_1| + (|z_2| + |z_3|)\,,
\end{aligned}$$

thus establishing the general result for three terms. The full inequality for n terms is demonstrated similarly.

The Cauchy–Schwarz Inequality: If $z_1, \ldots, z_n$ and $w_1, \ldots, w_n$ are complex numbers, then
$$\left| \sum_{j=1}^{n} z_j w_j \right|^2 \leq \left[\sum_{j=1}^{n} |z_j|^2 \right] \cdot \left[\sum_{j=1}^{n} |w_j|^2 \right]\,. \tag{2.34}$$

To understand why this inequality is true, let us begin with some special cases. For just one summand, the inequality says that
$$|z_1 w_1|^2 \leq |z_1|^2 |w_1|^2\,,$$

which is clearly true. For two summands, the inequality asserts that
$$|z_1 w_1 + z_2 w_2|^2 \leq (|z_1|^2 + |z_2|^2) \cdot (|w_1|^2 + |w_2|^2)\,.$$

Multiplying this out yields
$$|z_1 w_1|^2 + 2\mathrm{Re}\,(z_1 w_1 \overline{z_2 w_2}) + |z_2 w_2|^2 \leq |z_1|^2 |w_1|^2 + |z_1|^2 |w_2|^2 + |z_2|^2 |w_1|^2 + |z_2|^2 |w_2|^2\,.$$

Cancelling like terms, we have
$$2\mathrm{Re}\,(z_1 w_1 \overline{z_2 w_2}) \leq |z_1|^2 |w_2|^2 + |z_2|^2 |w_1|^2\,.$$

But it is always true, for $a, b \geq 0$, that $2ab \leq a^2 + b^2$. Hence
$$2\mathrm{Re}\,(z_1 w_1 \overline{z_2 w_2}) \leq 2|z_1 w_2||z_2 w_1| \leq |z_1 w_2|^2 + |z_2 w_1|^2\,.$$

The result for n terms is demonstrated similarly.

Exercises

1.   Find all the third roots of i.

2.   Find all the sixth roots of -1.

3.   Find all the fourth roots of $-i$.

4.   Find all the fifth roots of $-1 + i$.

5.   Find all arguments of each of these complex numbers: i, $1+i$, $-1+i\sqrt{3}$, $-2 - 2i$, $\sqrt{3} - i$.

6.   If z is any complex number then explain why

$$|z| \leq |\operatorname{Re} z| + |\operatorname{Im} z|.$$

7.   If z, w are any complex numbers then explain why

$$|z + w| \geq |z| - |w|.$$

8.   If $\sum_j |z_j|^2 < \infty$ and $\sum_j |w_j|^2 < \infty$ then explain why $\sum_j |z_j w_j| < \infty$.

Chapter 3

Holomorphic and Harmonic Functions

3.1 Holomorphic Functions

3.1.1 Continuously Differentiable and C^k Functions

In this book we will frequently refer to a *domain* or a *region* $U \subseteq \mathbb{C}$. Usually this will mean that U is an open set and that U is connected (see §§1.2.2).

Holomorphic functions are a generalization of complex polynomials. But they are more flexible objects than polynomials. The collection of all polynomials is closed under addition and multiplication. However, the collection of all holomorphic functions is closed under reciprocals, inverses, exponentiation, logarithms, square roots, and many other operations as well.

There are several different ways to introduce the concept of holomorphic function. They can be defined by way of power series, or using the complex derivative, or using partial differential equations. We shall touch on all these approaches; but our initial definition will be by way of partial differential equations.

If $U \subseteq \mathbb{R}^2$ is a region and $f : U \to \mathbb{R}$ is a continuous function, then f is called C^1 (or *continuously differentiable*) on U if $\partial f / \partial x$ and $\partial f / \partial y$ exist and are *continuous* on U. We write $f \in C^1(U)$ for short.

More generally, if $k \in \{0, 1, 2, ...\}$, then a real-valued function f on U is called C^k (k times continuously differentiable) if all partial derivatives of f up to and including order k exist and are continuous on U. We write in this case $f \in C^k(U)$. In particular, a C^0 function is just a continuous function.

EXAMPLE 1 Let $D \subseteq \mathbb{C}$ be the unit disc, $D = \{z \in \mathbb{C} : |z| < 1\}$. The function $\varphi(z) = |z|^2 = x^2 + y^2$ is C^k for every k. This is so just because we may differentiate φ as many times as we please, and the result is continuous. In this circumstance we sometimes write $\varphi \in C^\infty$.

By contrast, the function $\psi(z) = |z|$ is not even C^1. For the restriction of ψ to the real axis is $\widetilde{\psi}(x) = |x|$, and this function is well known not to be differentiable. □

A function $f = u + iv : U \to \mathbb{C}$ is called C^k if both u and v are C^k.

3.1.2 The Cauchy–Riemann Equations

If f is *any* complex-valued function, then we may write $f = u + iv$, where u and v are real-valued functions.

EXAMPLE 2 Consider

$$f(z) = z^2 = (x^2 - y^2) + i(2xy); \tag{3.1}$$

in this example $u = x^2 - y^2$ and $v = 2xy$. We refer to u as the *real part* of f and denote it by $\operatorname{Re} f$; we refer to v as the *imaginary part* of f and denote it by $\operatorname{Im} f$. □

Now we formulate the notion of "holomorphic function" in terms of the real and imaginary parts of f :

Let $U \subseteq \mathbb{C}$ be a region and $f : U \to \mathbb{C}$ a C^1 function. Write

$$f(z) = u(x, y) + iv(x, y), \tag{3.2}$$

with u and v real-valued functions. Of course $z = x + iy$ as usual. If u and v satisfy the equations

$$\frac{\partial u}{\partial x} = \frac{\partial v}{\partial y} \qquad \frac{\partial u}{\partial y} = -\frac{\partial v}{\partial x} \tag{3.3}$$

at every point of U, then the function f is said to be *holomorphic* (see §§3.1.4, where a more formal definition of "holomorphic" is provided). The first order, linear partial differential equations in (3.3) are called the *Cauchy–Riemann equations*. A practical method for checking whether a given function is holomorphic is to check whether it satisfies the Cauchy–Riemann equations. Another practical method is to check that the function can be expressed in terms of z alone, with no $\bar{z}$'s present (see §§3.1.3).

EXAMPLE 3 Let $f(z) = z^2 - z$. Then we may write

$$f(z) = (x+iy)^2 - (x+iy) = (x^2 - y^2 - x) + i(2xy - y) \equiv u(x,y) + iv(x,y).$$

Then we may check directly that

$$\frac{\partial u}{\partial x} = 2x - 1 = \frac{\partial v}{\partial y}$$

and

$$\frac{\partial u}{\partial y} = -2y = -\frac{\partial v}{\partial x}.$$

We see, then, that f satisfies the Cauchy–Riemann equations so it is holomorphic. Also observe that f may be expressed in terms of z alone, with no $\bar{z}$s. □

EXAMPLE 4 Define

$$\begin{aligned} g(z) &= |z|^2 - 4z + 2\bar{z} \\ &= z \cdot \bar{z} - 4z + 2\bar{z} \\ &= (x+iy) \cdot (x-iy) - 4 \cdot (x+iy) + 2(x-iy) \\ &= (x^2 + y^2 - 2x) + i(-6y) \\ &\equiv u(x,y) + iv(x,y). \end{aligned}$$

Then

$$\frac{\partial u}{\partial x} = 2x - 2 \neq -6 = \frac{\partial v}{\partial y}.$$

Also

$$\frac{\partial u}{\partial y} = 2y \neq 0 = -\frac{\partial v}{\partial x}.$$

We see that *both* Cauchy–Riemann equations fail. So g is not holomorphic. We may also observe that g is expressed both in terms of z and $\bar{z}$—another sure indicator that this function is not holomorphic. □

3.1.3   Derivatives

We define, for $f = u + iv : U \to \mathbb{C}$ a C^1 function,

$$\frac{\partial}{\partial z}f \equiv \frac{1}{2}\left(\frac{\partial}{\partial x} - i\frac{\partial}{\partial y}\right)f = \frac{1}{2}\left(\frac{\partial u}{\partial x} + \frac{\partial v}{\partial y}\right) + \frac{i}{2}\left(\frac{\partial v}{\partial x} - \frac{\partial u}{\partial y}\right) \qquad (3.4)$$

and

$$\frac{\partial}{\partial \bar{z}}f \equiv \frac{1}{2}\left(\frac{\partial}{\partial x} + i\frac{\partial}{\partial y}\right)f = \frac{1}{2}\left(\frac{\partial u}{\partial x} - \frac{\partial v}{\partial y}\right) + \frac{i}{2}\left(\frac{\partial v}{\partial x} + \frac{\partial u}{\partial y}\right). \qquad (3.5)$$

If $z = x + iy$, $\bar{z} = x - iy$, then one can check directly that

$$\frac{\partial}{\partial z}z = 1, \qquad \frac{\partial}{\partial z}\bar{z} = 0, \qquad (3.6)$$

$$\frac{\partial}{\partial \bar{z}}z = 0, \qquad \frac{\partial}{\partial \bar{z}}\bar{z} = 1. \qquad (3.7)$$

If a C^1 function f satisfies $\partial f/\partial z \equiv 0$ on an open set U, then f does not depend on z (but it *can* depend on $\bar{z}$). If instead f satisfies $\partial f/\partial \bar{z} \equiv 0$ on an open set U, then f does not depend on $\bar{z}$ (but it *does* depend on z). The condition $\partial f/\partial \bar{z} = 0$ is just a reformulation of the Cauchy–Riemann equations—see §§3.1.2. We work out the details of this claim in §§3.1.4. Now we look at some examples to illustrate the new ideas.

EXAMPLE 5   Review Example 3. Now let us examine that same function using our new criterion with the operator $\partial/\partial \bar{z}$. We have

$$\frac{\partial}{\partial \bar{z}}f(z) = \frac{\partial}{\partial \bar{z}}\left(z^2 - z\right) = 2z\frac{\partial z}{\partial \bar{z}} - \frac{\partial z}{\partial \bar{z}} = 0 - 0 = 0.$$

We conclude that f is holomorphic.   □

EXAMPLE 6   Review Example 4. Now let us examine that same function using our new criterion with the operator $\partial/\partial \bar{z}$. We have

$$\frac{\partial}{\partial \bar{z}}g(z) = \frac{\partial}{\partial \bar{z}}\left(|z|^2 - 4z + 2\bar{z}\right) = \frac{\partial}{\partial \bar{z}}\left(z \cdot \bar{z} - 4z + 2\bar{z}\right) = z + 2 \neq 0.$$

We conclude that g is *not* holomorphic.   □

3.1.4 Definition of Holomorphic Function

Functions f that satisfy $(\partial/\partial\bar{z})f \equiv 0$ are the main concern of complex analysis. A continuously differentiable (C^1) function $f : U \to \mathbb{C}$ defined on an open subset U of $\mathbb{C}$ is said to be *holomorphic* if

$$\frac{\partial f}{\partial \bar{z}} = 0 \qquad (3.8)$$

at every point of U. Note that this last equation is just a reformulation of the Cauchy–Riemann equations (§§3.1.2). To see this, we calculate:

$$
\begin{aligned}
0 &= \frac{\partial}{\partial \bar{z}} f(z) \\
&= \frac{1}{2}\left(\frac{\partial}{\partial x} + i\frac{\partial}{\partial y}\right)[u(z) + iv(z)] \\
&= \left[\frac{\partial u}{\partial x} - \frac{\partial v}{\partial y}\right] + i\left[\frac{\partial u}{\partial y} + \frac{\partial v}{\partial x}\right].
\end{aligned}
\qquad (3.9)
$$

Of course the far right-hand side cannot be identically zero unless each of its real and imaginary parts is identically zero. It follows that

$$\frac{\partial u}{\partial x} - \frac{\partial v}{\partial y} = 0 \qquad (3.10)$$

and

$$\frac{\partial u}{\partial y} + \frac{\partial v}{\partial x} = 0. \qquad (3.11)$$

These are the Cauchy–Riemann equations (3.3).

EXAMPLE 7 The function $h(z) = z^3 - 4z^2 + z$ is holomorphic because

$$\frac{\partial}{\partial \bar{z}} h(z) = 3z^2 \frac{\partial z}{\partial \bar{z}} - 4 \cdot 2z \frac{\partial z}{\partial \bar{z}} + \frac{\partial z}{\partial \bar{z}} = 0. \qquad \square$$

3.1.5 Examples of Holomorphic Functions

Certainly any polynomial in z (*without* $\bar{z}$) is holomorphic. And the reciprocal of any polynomial is holomorphic, as long as we restrict attention to a region where the polynomial does not vanish.

Earlier in this book we have discussed the complex function

$$e^z = \sum_{j=0}^{\infty} \frac{z^j}{j!} \,.$$

One may calculate directly, just differentiating the power series term-by-term, that

$$\frac{\partial}{\partial z} e^z = e^z \,.$$

In addition,

$$\frac{\partial}{\partial \bar{z}} e^z = 0 \,,$$

so the exponential function is holomorphic.

3.1.6   The Complex Derivative

Let $U \subseteq \mathbb{C}$ be open, $P \in U$, and $g : U \setminus \{P\} \to \mathbb{C}$ a function. We say that

$$\lim_{z \to P} g(z) = \ell \,, \quad \ell \in \mathbb{C} \,, \tag{3.12}$$

if, for any $\epsilon > 0$ there is a $\delta > 0$ such that when $z \in U$ and $0 < |z - P| < \delta$ then $|g(z) - \ell| < \epsilon$. Notice that, in this definition of limit, the point z may approach P in an arbitrary manner—from any direction. See Figure 2.1. Of course the function g is *continuous* at $P \in U$ if $\lim_{z \to P} g(z) = g(P)$.

We say that f possesses the *complex derivative* at P if

$$\lim_{z \to P} \frac{f(z) - f(P)}{z - P} \tag{3.13}$$

exists. In that case we denote the limit by $f'(P)$ or sometimes by

$$\frac{df}{dz}(P) \quad \text{or} \quad \frac{\partial f}{\partial z}(P). \tag{3.14}$$

This notation is consistent with that introduced in §§3.1.3: for a *holomorphic function*, the complex derivative calculated according to formula (3.13) or according to formula (3.4) is just the same. We shall say more about the complex derivative in §§3.2.1 and §§3.2.2.

We repeat that, in calculating the limit in (3.13), z must be allowed to approach P from *any* direction (refer to Figure 3.1). As an example,

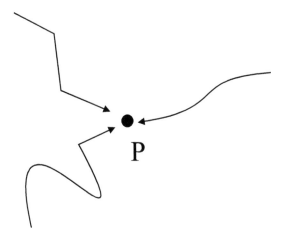

Figure 3.1: The point may approach P arbitrarily.

the function $g(x, y) = x - iy$—equivalently, $g(z) = \bar{z}$—does *not* possess the complex derivative at 0. To see this, calculate the limit

$$\lim_{z \to P} \frac{g(z) - g(P)}{z - P} \tag{3.15}$$

with z approaching $P = 0$ through values $z = x + i0$. The answer is

$$\lim_{x \to 0} \frac{x - 0}{x - 0} = 1. \tag{3.16}$$

If instead z is allowed to approach $P = 0$ through values $z = iy$, then the value is

$$\lim_{z \to P} \frac{g(z) - g(P)}{z - P} = \lim_{y \to 0} \frac{-iy - 0}{iy - 0} = -1. \tag{3.17}$$

Observe that the two answers do not agree. *In order for the complex derivative to exist, the limit must exist and assume only one value no matter how z approaches P.* Therefore, this example g does not possess the complex derivative at $P = 0$. In fact a similar calculation shows that this function g does *not* possess the complex derivative at any point.

If a function f possesses the complex derivative at every point of its open domain U, then f is holomorphic. This definition is equivalent to definitions given in §§3.1.4. We repeat some of these ideas in §3.2. In fact, from an historical perspective, it is important to recall a theorem of Goursat (see the Appendix in [GRK]). It says that if a function f

possesses the complex derivative at each point of an open region $U \subseteq \mathbb{C}$ then f is in fact continuously differentiable on U. One may then verify the Cauchy–Riemann equations, and it follows that f is holomorphic by any of our definitions thus far.

3.1.7   Alternative Terminology for Holomorphic Functions

Some books use the word "analytic" instead of "holomorphic." Still others say "differentiable" or "complex differentiable" instead of "holomorphic." The use of the term "analytic" derives from the fact that a holomorphic function has a local power series expansion about each point of its domain (see §§5.1.6). In fact this power series property is a complete characterization of holomorphic functions; we shall discuss it in detail below. The use of "differentiable" derives from properties related to the complex derivative. These pieces of terminology and their significance will all be sorted out as the book develops. Somewhat archaic terms for holomorphic functions, which may be found in older texts, are "regular" and "monogenic."

Another piece of terminology that is applied to holomorphic functions is "conformal" or "conformal mapping." "Conformality" is an important geometric property of holomorphic functions that make these functions useful for modeling incompressible fluid flow (§§12.2.2) and other physical phenomena. We shall discuss conformality in §§3.3.1 and Chapter 11. We shall treat physical applications of conformality in Chapter 12.

Exercises

1.   Verify that each of these functions is holomorphic whereever it is defined:

   (a) $f(z) = \sin z - \dfrac{z^2}{z+1}$

   (b) $g(z) = e^{2z - z^3} - z^2$

   (c) $h(z) = \dfrac{\cos z}{z^2 + 1}$

   (d) $k(z) = z(\sin z + z)$

2. Verify that each of these functions is *not* holomorphic:

(a) $f(z) = |z|^4 - |z|^2$

(b) $g(z) = \dfrac{\overline{z}}{z^2 + 1}$

(c) $h(z) = z(\overline{z}^2 - z)$

(d) $k(z) = \overline{z} \cdot (\sin z) \cdot (\cos \overline{z})$

3. For each function f, calculate $\partial f / \partial z$:

(a) $2z(1 - z^3)$

(b) $(\cos z) \cdot (1 + \sin^2 z)$

(c) $(\sin \overline{z})(1 + \overline{z} \cos z)$

(d) $|z|^4 - |z|^2$

4. For each function g, calculate $\partial g / \partial \overline{z}$:

(a) $2\overline{z}(1 - z^3)$

(b) $(\sin \overline{z}) \cdot (1 + \sin^2 \overline{z})$

(c) $(\cos z) \cdot (1 + z \cos \overline{z})$

(d) $|z|^2 - |z|^4$

5. Verify the equations

$$\frac{\partial}{\partial z} z = 1, \qquad \frac{\partial}{\partial z} \overline{z} = 0,$$

$$\frac{\partial}{\partial \overline{z}} z = 0, \qquad \frac{\partial}{\partial \overline{z}} \overline{z} = 1.$$

6. Calculate the derivative

$$\frac{\partial}{\partial z}[\cos z - e^{3z}].$$

7. Calculate the derivative

$$\frac{\partial}{\partial \overline{z}}[\sin \overline{z} - z\overline{z}^2].$$

8.   Find a function g such that

$$\frac{\partial g}{\partial z} = z\bar{z}^2 - \sin z \,.$$

9.   Find a function h such that

$$\frac{\partial h}{\partial \bar{z}} = \bar{z}^2 z^3 + \cos \bar{z} \,.$$

10.   Find a function k such that

$$\frac{\partial^2 k}{\partial z \partial \bar{z}} = |z|^2 - \sin z + \bar{z}^3 \,.$$

11.   From the definition (line (3.13)), calculate

$$\frac{d}{dz}(z^3 - z^2) \,.$$

12.   From the definition (line (3.13)), calculate

$$\frac{d}{dz}(\sin z - e^z) \,.$$

3.2   The Relationship of Holomorphic and Harmonic Functions

3.2.1   Harmonic Functions

A C^2 function u is said to be *harmonic* if it satisfies the equation

$$\left(\frac{\partial^2}{\partial x^2} + \frac{\partial^2}{\partial y^2}\right) u = 0. \tag{3.18}$$

This equation is called *Laplace's equation,* and is frequently abbreviated as

$$\triangle u = 0. \tag{3.19}$$

EXAMPLE 8 The function $u(x, y) = x^2 - y^2$ is harmonic. This assertion may be verified directly:

$$\triangle u = \left(\frac{\partial^2}{\partial x^2} + \frac{\partial^2}{\partial y^2}\right) u = \left(\frac{\partial^2}{\partial x^2}\right) x^2 - \left(\frac{\partial^2}{\partial y^2}\right) y^2 = 2 - 2 = 0.$$

A similar calculation shows that $v(x, y) = 2xy$ is harmonic. For

$$\triangle v = \left(\frac{\partial^2}{\partial x^2} + \frac{\partial^2}{\partial y^2}\right) 2xy = 0 + 0 = 0. \qquad \square$$

EXAMPLE 9 The function $\tilde{u}(x, y) = x^3$ is *not* harmonic. For

$$\triangle \tilde{u} = \left(\frac{\partial^2}{\partial x^2} + \frac{\partial^2}{\partial y^2}\right) u = \left(\frac{\partial^2}{\partial x^2} + \frac{\partial^2}{\partial y^2}\right) x^3 = 6x \neq 0.$$

Likewise, the function $\tilde{v}(x, y) = \sin x - \cos y$ is *not* harmonic. For

$$\triangle \tilde{v} = \left(\frac{\partial^2}{\partial x^2} + \frac{\partial^2}{\partial y^2}\right) \tilde{v} = \left(\frac{\partial^2}{\partial x^2} + \frac{\partial^2}{\partial y^2}\right) [\sin x - \cos y] = -\sin x + \cos y \neq 0.$$

$\square$

$\square$

3.2.2 Holomorphic and Harmonic Functions

If f is a holomorphic function and $f = u + iv$ is the expression of f in terms of its real and imaginary parts, then both u and v are harmonic. The easiest way to see this is to begin with the equation

$$\frac{\partial}{\partial \bar{z}} f = 0 \qquad (3.20)$$

and to apply $\partial/\partial z$ to both sides. The result is

$$\frac{\partial}{\partial z} \frac{\partial}{\partial \bar{z}} f = 0 \qquad (3.21)$$

or

$$\left(\frac{1}{2}\left[\frac{\partial}{\partial x} - i\frac{\partial}{\partial y}\right]\right) \left(\frac{1}{2}\left[\frac{\partial}{\partial x} + i\frac{\partial}{\partial y}\right]\right) [u + iv] = 0. \qquad (3.22)$$

Multiplying through by 4, and then multiplying out the derivatives, we find that

$$\left(\frac{\partial^2}{\partial x^2} + \frac{\partial^2}{\partial y^2}\right)[u + iv] = 0. \qquad (3.23)$$

We may now distribute the differentiation and write this as

$$\left(\frac{\partial^2}{\partial x^2} + \frac{\partial^2}{\partial y^2}\right)u \; + \; i\left(\frac{\partial^2}{\partial x^2} + \frac{\partial^2}{\partial y^2}\right)v = 0. \qquad (3.24)$$

The only way that the left-hand side can be zero is if its real part is zero and its imaginary part is zero. We conclude then that

$$\left(\frac{\partial^2}{\partial x^2} + \frac{\partial^2}{\partial y^2}\right)u = 0 \qquad (3.25)$$

and

$$\left(\frac{\partial^2}{\partial x^2} + \frac{\partial^2}{\partial y^2}\right)v = 0. \qquad (3.26)$$

Thus u and v are each harmonic.

EXAMPLE 10 Let $f(z) = (z + z^2)^2$. Then f is certainly holomorphic because it is defined using only zs, and no $\bar{z}$s. Notice that

$$\begin{aligned} f(z) \;&=\; z^4 + 2z^3 + z^2 \\ &=\; [x^4 - 6x^2y^2 + y^4 + 2x^3 - 6xy^2 + x^2 - y^2] \\ &\quad + i[-4xy^3 + 4x^3y + 6x^2y - 2y^3 + 2xy] \\ &\equiv\; u + iv. \end{aligned}$$

We may check directly that

$$\triangle u = 0 \qquad \text{and} \qquad \triangle v = 0.$$

Hence the real and imaginary parts of f are each harmonic.   □

A sort of converse to (3.25) and (3.26) is true provided the functions involved are defined on a domain with no holes:

If $\mathcal{R}$ is an open rectangle (or open disc) and if u is a real-valued harmonic function on $\mathcal{R}$, then there is a holomorphic function F on $\mathcal{R}$ such that $\operatorname{Re} F = u$. In other words, for such a function u there exists another harmonic function v defined on $\mathcal{R}$ such that $F \equiv u + iv$ is holomorphic on $\mathcal{R}$. Any two such functions v must differ by a real constant.

More generally, if U is a region with no holes (a *simply connected* region—see §§4.2.2), and if u is harmonic on U, then there is a holomorphic function F on U with $\operatorname{Re} F = u$. In other words, for such a function u there exists a harmonic function v defined on U such that $F \equiv u + iv$ is holomorphic on U. Any two such functions v must differ by a constant. We call the function v a *harmonic conjugate* for u.

The displayed statement is false on a domain with a hole, such as an annulus. For example, the harmonic function $u = \log(x^2 + y^2)$, defined on the annulus $U = \{z : 1 < |z| < 2\}$, has no harmonic conjugate on U. See also §§3.2.2.

EXAMPLE 11 Consider the function $u(x, y) = x^2 - y^2 - x$ on the square $U = \{(x, y) : |x| < 1, |y| < 1\}$. Certainly U is simply connected. And one may verify directly that $\triangle u \equiv 0$ on U. To solve for v a harmonic conjugate of u, we use the Cauchy–Riemann equations:

$$\frac{\partial v}{\partial y} = \frac{\partial u}{\partial x} = 2x - 1,$$
$$\frac{\partial v}{\partial x} = -\frac{\partial u}{\partial y} = 2y.$$

The first of these equations indicates that $v(x, y) = 2xy - y + \varphi(x)$, for some unknown function $\varphi(x)$. Then

$$2y = \frac{\partial v}{\partial x} = 2y - \varphi'(x).$$

It follows that $\varphi'(x) = 0$ so that $\varphi(x) \equiv C$ for some real constant C. In conclusion,

$$v(x, y) = 2xy - y + C.$$

In other words, $h(x, y) = u(x, y) + iv(x, y) = [x^2 - y^2 - x] + i[2xy - y + C]$ should be holomorphic. We may verify this claim immediately by writing h as

$$h(z) = z^2 - z + iC.$$ □

You may also verify that the function h in the last example is holomorphic by checking the Cauchy–Riemann equations.

Exercises

1. Verify that each of these functions is harmonic:

   (a) $f(z) = \operatorname{Re} z$

   (b) $g(z) = x^3 - 3xy^2$

   (c) $h(z) = |z|^2 - 2x^2$

   (d) $k(z) = e^x \cos y$

2. Verify that each of these functions is *not* harmonic:

   (a) $f(z) = |z|^2$

   (b) $g(z) = |z|^4 - |z|^2$

   (c) $h(z) = \bar{z} \sin z$

   (d) $k(z) = e^{\bar{z} \cos z}$

3. For each of these (real-valued) harmonic functions u, find a (real-valued) harmonic function v such that $u + iv$ is holomorphic.

   (a) $u(z) = e^x \sin y$

   (b) $u(z) = 3x^2 y - y^3$

   (c) $u(z) = e^{2y} \sin x \cos x$

   (d) $u(z) = x - y$

4. Use the chain rule to express the Laplace operator $\triangle$ in terms of polar coordinates (r, θ).

5. Let $\rho(x, y)$ be a rotation of the plane. Thus ρ is given by a 2×2 matrix with each row a unit vector and the two rows orthogonal to each other. Further, the determinant of the matrix is 1. Show that, for any C^2 function f,

$$\triangle(f \circ \rho) = (\triangle f) \circ \rho.$$

6. Let $a \in \mathbb{R}^2$ and let λ_a be the operator $\lambda_a(x, y) = (x, y) + a$. This is translation by a. Verify that, for any C^2 function f,

$$\triangle(f \circ \lambda_a) = (\triangle f) \circ \lambda_a.$$

7. A function u is *biharmonic* if $\triangle^2 u = 0$. Verify that the function $x^4 - y^4$ is biharmonic. Give two distinct other examples of non-constant biharmonic functions. [Note that biharmonic functions are useful in the study of charge-transfer reactions in physics.]

8. Calculate the real and imaginary parts of the holomorphic function

$$f(z) = z^2 \cos z - e^{z^3 - z}$$

and verify directly that each of these functions is harmonic.

3.3 Complex Differentiability and Conformality

3.3.1 Conformality

Now we make some remarks about "conformality." Stated loosely, a function is *conformal* at a point $P \in \mathbb{C}$ if the function "preserves angles" at P and "stretches equally in all directions" at P. Both of these statements must be interpreted infinitesimally; we shall learn to do so in the discussion below. Holomorphic functions enjoy both properties:

Let f be holomorphic in a neighborhood of $P \in \mathbb{C}$. Let w_1, w_2 be complex numbers of unit modulus. Consider the directional derivatives

$$D_{w_1} f(P) \equiv \lim_{t \to 0} \frac{f(P + tw_1) - f(P)}{t} \qquad (3.27)$$

and

$$D_{w_2} f(P) \equiv \lim_{t \to 0} \frac{f(P + t w_2) - f(P)}{t}. \tag{3.28}$$

Then

(3.29) $|D_{w_1} f(P)| = |D_{w_2} f(P)|$.

(3.30) If $|f'(P)| \neq 0$, then the directed angle from w_1 to w_2 equals the directed angle from $D_{w_1} f(P)$ to $D_{w_2} f(P)$.

Statement **(3.29)** is the analytical formulation of "stretching equally in all directions." Statement **(3.30)** is the analytical formulation of "preserves angles."

In fact let us now give a discursive description of why conformality works. Either of these two properties actually characterizes holomorphic functions. It is worthwhile to picture the matter in the following manner: Let f be holomorphic on the open set $U \subseteq \mathbb{C}$. Fix a point $P \in U$. Write $f = u + iv$ as usual. Thus we may write the mapping f as $(x, y) \mapsto (u, v)$. Then the (real) Jacobian matrix of the mapping is

$$J(P) = \begin{pmatrix} u_x(P) & u_y(P) \\ v_x(P) & v_y(P) \end{pmatrix},$$

where subscripts denote derivatives. We may use the Cauchy–Riemann equations to rewrite this matrix as

$$J(P) = \begin{pmatrix} u_x(P) & u_y(P) \\ -u_y(P) & u_x(P) \end{pmatrix}$$

Factoring out a numerical coefficient, we finally write this two-dimensional derivative as

$$
\begin{aligned}
J(P) \;=\; & \sqrt{u_x(P)^2 + u_y(P)^2} \\
& \cdot \begin{pmatrix} \dfrac{u_x(P)}{\sqrt{u_x(P)^2 + u_y(P)^2}} & \dfrac{u_y(P)}{\sqrt{u_x(P)^2 + u_y(P)^2}} \\[2ex] \dfrac{-u_y(P)}{\sqrt{u_x(P)^2 + u_y(P)^2}} & \dfrac{u_x(P)}{\sqrt{u_x(P)^2 + u_y(P)^2}} \end{pmatrix} \\[2ex]
\equiv \; & h(P) \cdot \mathcal{J}(P).
\end{aligned}
$$

The matrix $\mathcal{J}(P)$ is of course a special orthogonal matrix (i.e., its rows form an orthonormal basis of $\mathbb{R}^2$, and it is oriented positively). Of course a special orthogonal matrix represents a *rotation*. Thus we see that the derivative of our mapping is a rotation $\mathcal{J}(P)$ (which preserves angles) followed by a positive "stretching factor" $h(P)$ (which also preserves angles).

In fact the second characterization of conformality (in terms of preservation of directed angles) has an important converse: If **(3.30)** holds at points near P, then f has a complex derivative at P. As for the first characterization of conformality, if **(3.29)** holds at points near P, then either f or $\overline{f}$ has a complex derivative at P.

A function that is conformal at all points of an open set U must possess the complex derivative at each point of U. By the discussion in §§3.1.6, the function f is therefore holomorphic if it is C^1. Or, by Goursat's theorem, it would then follow that the function is holomorphic on U, with the C^1 condition being automatic.

Exercises

1. Consider the holomorphic function $f(z) = z^2$. Calculate the derivative of f at the point $P = 1 + i$. Write down the Jacobian matrix of f at P, thought of as a 2×2 real matrix operator. Verify directly (by imitating the calculations presented in this section) that this Jacobian matrix is the composition of a special orthogonal matrix and a dilation.

2. Repeat the first exercise with the function $g(z) = \sin z$ and $P = \pi + (\pi/2)i$.

3. Repeat the first exercise with the function $h(z) = e^z$ and $P = 2 - i$.

4. Discuss, in physical language, why the surface motion of an incompressible fluid flow should be conformal.

5. Verify that the function $g(z) = \overline{z}^2$ has the property that (at all points not equal to 0) it stretches equally in all directions, but it reverses angles. We say that such a function is *anticonformal*.

6. The function $h(z) = z + 2\overline{z}$ is *not* conformal. Explain why.

7.   If a continuously differentiable function is conformal then it is holomorphic. Explain why.

8.   If f is conformal then any positive integer power of f is conformal. Explain why.

9.   If f is conformal then e^f is conformal. Explain why.

10.   Let $\Omega \subseteq \mathbb{C}$ be a domain and $\varphi : \Omega \to \mathbb{R}$ be a function. Explain why φ, no matter how smooth or otherwise well behaved, could not possibly be conformal.

Chapter 4

The Cauchy Theory

4.1   Real and Complex Line Integrals

In this section we shall recast the line integral from calculus in complex notation. The result will be the complex line integral.

4.1.1   Curves

It is convenient to think of a *curve* as a (continuous) function γ from a closed interval $[a, b] \subseteq \mathbb{R}$ into $\mathbb{R}^2 \approx \mathbb{C}$. In practice it is useful *not* to distinguish between the *function* γ and the image (or set of points that make up the curve) given by $\{\gamma(t) : t \in [a, b]\}$. In the case that $\gamma(a) = \gamma(b)$, then we say that the curve is *closed*. Refer to Figure 4.1.

It is often convenient to write

$$\gamma(t) = (\gamma_1(t), \gamma_2(t)) \qquad \text{or} \qquad \gamma(t) = \gamma_1(t) + i\gamma_2(t). \qquad (4.1)$$

For example, $\gamma(t) = (\cos t, \sin t) = \cos t + i \sin t$, $t \in [0, 2\pi]$, describes the unit circle in the plane. The circle is traversed in a counterclockwise manner as t increases from 0 to 2π. This curve is closed. Refer to Figure 4.2.

4.1.2   Closed Curves

We have already noted that the curve $\gamma : [a, b] \to \mathbb{C}$ is called *closed* if $\gamma(a) = \gamma(b)$. It is called *simple, closed* (or Jordan) if the restriction of

45

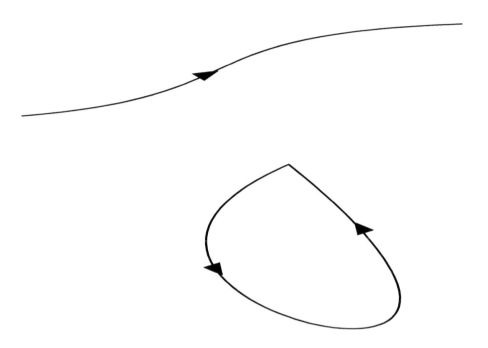

Figure 4.1: Two curves in the plane, one closed.

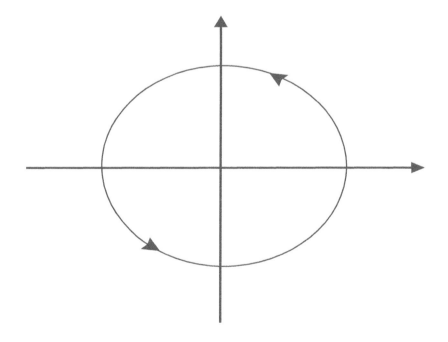

Figure 4.2: A simple, closed curve (the unit circle).

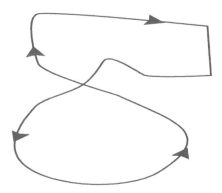

Figure 4.3: A closed curve that is not simple.

γ to the interval $[a, b)$ (which is commonly written $\gamma|_{[a,b)}$) is one-to-one and $\gamma(a) = \gamma(b)$ (Figures 4.2, 4.3). Intuitively, a simple, closed curve is a curve with no self-intersections, except of course for the closing up at $t = a, b$.

In order to work effectively with γ we need to impose on it some differentiability properties.

4.1.3 Differentiable and C^k Curves

A function $\varphi : [a, b] \to \mathbb{R}$ is called *continuously differentiable* (or C^1), and we write $\varphi \in C^1([a, b])$, if

(4.2) φ is continuous on $[a, b]$;

(4.3) φ' exists on (a, b);

(4.4) φ' has a continuous extension to $[a, b]$.

In other words, we require that

$$\lim_{t \to a^+} \varphi'(t) \quad \text{and} \quad \lim_{t \to b^-} \varphi'(t) \tag{4.5}$$

both exist.

Note that, under these circumstances,

$$\varphi(b) - \varphi(a) = \int_a^b \varphi'(t)\, dt, \tag{4.6}$$

so that the Fundamental Theorem of Calculus holds for $\varphi \in C^1([a, b])$.

A curve $\gamma : [a, b] \to \mathbb{C}$, with $\gamma(t) = \gamma_1(t) + i\gamma_2(t)$ is said to be *continuous* on $[a, b]$ if both γ_1 and γ_2 are. The curve is *continuously differentiable* (or C^1) on $[a, b]$, and we write

$$\gamma \in C^1([a, b]), \tag{4.7}$$

if γ_1, γ_2 are continuously differentiable on $[a, b]$. Under these circumstances we will write

$$\frac{d\gamma}{dt} = \frac{d\gamma_1}{dt} + i\frac{d\gamma_2}{dt}. \tag{4.8}$$

We also sometimes write $\gamma'(t)$ or $\dot{\gamma}(t)$ for $d\gamma/dt$.

4.1.4   Integrals on Curves

Let $\psi : [a, b] \to \mathbb{C}$ be continuous on $[a, b]$. Write $\psi(t) = \psi_1(t) + i\psi_2(t)$. Then we define

$$\int_a^b \psi(t)\, dt \equiv \int_a^b \psi_1(t)\, dt + i \int_a^b \psi_2(t)\, dt \tag{4.9}$$

We summarize the ideas presented thus far by noting that, if $\gamma \in C^1([a, b])$ is complex-valued, then

$$\gamma(b) - \gamma(a) = \int_a^b \gamma'(t)\, dt. \tag{4.10}$$

4.1.5   The Fundamental Theorem of Calculus along Curves

Now we state the Fundamental Theorem of Calculus (see [THO], [BLK]) along curves.

Let $U \subseteq \mathbb{C}$ be a domain and let $\gamma : [a, b] \to U$ be a C^1 curve. If $f \in C^1(U)$, then

$$f(\gamma(b)) - f(\gamma(a)) = \int_a^b \left(\frac{\partial f}{\partial x}(\gamma(t)) \cdot \frac{d\gamma_1}{dt} + \frac{\partial f}{\partial y}(\gamma(t)) \cdot \frac{d\gamma_2}{dt} \right) dt. \tag{4.11}$$

4.1.6 The Complex Line Integral

When f is holomorphic, then formula (4.11) may be rewritten (using the Cauchy–Riemann equations) as

$$f(\gamma(b)) - f(\gamma(a)) = \int_a^b \frac{\partial f}{\partial z}(\gamma(t)) \cdot \frac{d\gamma}{dt}(t)\, dt, \qquad (4.12)$$

where, as earlier, we have taken $d\gamma/dt$ to be $d\gamma_1/dt + id\gamma_2/dt$.

This latter result plays much the same role for holomorphic functions as does the Fundamental Theorem of Calculus for functions from $\mathbb{R}$ to $\mathbb{R}$. The expression on the right of (4.12) is called the *complex line integral* and is denoted

$$\oint_\gamma \frac{\partial f}{\partial z}(z)\, dz. \qquad (4.13)$$

The small circle through the integral sign $\int$ tells us that this is a complex line integral, and has the meaning (4.12).

EXAMPLE 1 Let $f(z) = z^2 - 2z$ and let $\gamma(t) = (\cos t, \sin t)$, $0 \le t \le \pi$. This curve γ traverses the upper half of the unit circle from the initial point $(1, 0)$ to the terminal point $(-1, 0)$. We may calculate that

$$\begin{aligned}
\oint_\gamma f(z)\, dz &= \int_0^\pi f(\cos t + i \sin t) \cdot (-\sin t + i \cos t)\, dt \\
&= \int_0^\pi \left[(\cos t + i \sin t)^2 - 2(\cos t + i \sin t)\right] \cdot (-\sin t + i \cos t)\, dt \\
&= \int_0^\pi 4\cos t \sin t - 3\sin t \cos^2 t - 2i \cos 2t - 3i \sin^2 t \cos t \\
&\qquad + \sin^3 t + i \cos^3 t\, dt \\
&= \Big[2\sin^2 t + \cos^3 t - i\sin 2t - i\sin^3 t - \cos t + i \sin t \\
&\qquad + \frac{\cos^3 t}{3} - i\frac{\sin^3 t}{3} \Big]_0^\pi \\
&= -\frac{2}{3}. \qquad\qquad\qquad\qquad\qquad\qquad\qquad\qquad \square
\end{aligned}$$

EXAMPLE 2 If we integrate the holomorphic function f from the last example around the closed curve $\eta(t) = (\cos t, \sin t)$, $0 \leq t \leq 2\pi$, then we obtain

$$
\begin{aligned}
\oint_\eta f(z)\, dz &= \int_0^{2\pi} f(\cos t + i \sin t) \cdot (-\sin t + i \cos t)\, dt \\
&= \int_0^{2\pi} \left[(\cos t + i \sin t)^2 - 2(\cos t + i \sin t) \right] \cdot (-\sin t + i \cos t)\, dt \\
&= \int_0^{2\pi} 4 \cos t \sin t - 3 \sin t \cos^2 t - 2i \cos 2t - 3i \sin^2 t \cos t \\
&\qquad + \sin^3 t + i \cos^3 t\, dt \\
&= \left[2 \sin^2 t + \cos^3 t - i \sin 2t - i \sin^3 t - \cos t + i \sin t \right. \\
&\qquad \left. + \frac{\cos^3 t}{3} - \frac{\sin^3 t}{3} \right]_0^{2\pi} \\
&= 0\,.   \hspace{8cm} \square
\end{aligned}
$$

More generally, if g is *any* continuous function whose domain contains the curve γ, then the complex line integral of g along γ is defined to be

$$
\oint_\gamma g(z)\, dz \equiv \int_a^b g(\gamma(t)) \cdot \frac{d\gamma}{dt}(t)\, dt. \tag{4.14}
$$

The whole concept of complex line integral is central to our further considerations in later sections. We shall use integrals like the one on the right of (4.12) or (4.14) even when f is not holomorphic; but we can be sure that the equality (4.12) holds only when f *is* holomorphic.

EXAMPLE 3 Let $g(z) = |z|^2$ and let $\mu(t) = t + it$, $0 \leq t \leq 1$. Let us calculate

$$
\oint_\mu g(z)\, dz\,.
$$

We have

$$
\oint_\mu g(z)\, dz = \int_0^1 g(t+it) \cdot \mu'(t)\, dt = \int_0^1 2t^2 \cdot (1+i) dt = \left. \frac{2t^3}{3}(1+i) \right|_0^1 = \frac{2+2i}{3}\,.
$$

$$\square$$

4.1.7 Properties of Integrals

We conclude this section with some easy but useful facts about integrals.

(4.15) If $\varphi : [a, b] \to \mathbb{C}$ is continuous, then

$$\left| \int_a^b \varphi(t)\, dt \right| \leq \int_a^b |\varphi(t)|\, dt.$$

(4.16) If $\gamma : [a, b] \to \mathbb{C}$ is a C^1 curve and φ is a continuous function on the curve γ, then

$$\left| \oint_\gamma \varphi(z)\, dz \right| \leq \left[\max_{t \in [a,b]} |\varphi(t)| \right] \cdot \ell(\gamma), \qquad (4.16a)$$

where

$$\ell(\gamma) \equiv \int_a^b |\varphi'(t)|\, dt$$

is the *length* of γ.

(4.17) The calculation of a complex line integral is independent of the way in which we parameterize the path:

> Let $U \subseteq \mathbb{C}$ be an open set and $F : U \to \mathbb{C}$ a continuous function. Let $\gamma : [a, b] \to U$ be a C^1 curve. Suppose that $\varphi : [c, d] \to [a, b]$ is a one-to-one, onto, increasing C^1 function with a C^1 inverse. Let $\widetilde{\gamma} = \gamma \circ \varphi$. Then
> $$\oint_{\widetilde{\gamma}} f\, dz = \oint_\gamma f\, dz.$$

This last statement implies that one can use the idea of the integral of a function f along a curve γ when the curve γ is described geometrically but without reference to a specific parameterization. For instance, "the integral of $\bar{z}$ *counterclockwise* around the unit circle $\{z \in \mathbb{C} : |z| = 1\}$" is now a phrase that makes sense, even though we have not indicated a specific parameterization of the unit circle. Note, however, that the direction counts: The integral of $\bar{z}$ counterclockwise around the unit circle is $2\pi i$. If the direction is reversed, then the integral changes sign: The integral of $\bar{z}$ *clockwise* around the unit circle is $-2\pi i$.

EXAMPLE 4 Let $g(z) = z^2 - z$ and $\gamma(t) = t^2 - it$, $0 \le t \le 1$. Then

$$\oint_\gamma g(z)\,dz = \int_0^1 g(t^2 - it) \cdot \gamma'(t)\,dt \;\; = \;\; \int_0^1 [(t^2 - it)^2 - (t^2 - it)] \cdot (2t - i)\,dt$$

$$= \int_0^1 [t^4 - 2it^3 - 2t^2 + it] \cdot (2t - i)\,dt$$

$$= \int_0^1 2t^5 - 5it^4 - 6t^3 + 4it^2 + t\,dt$$

$$= \left[\frac{2t^6}{6} - \frac{5it^5}{5} - \frac{6t^4}{4} + \frac{4it^3}{3} + \frac{t^2}{2} \right]_0^1$$

$$= \frac{1}{3} - i - \frac{3}{2} + \frac{4i}{3} + \frac{1}{2}$$

$$= -\frac{2}{3} + \frac{i}{3}\,.$$

If instead we replace γ by $-\gamma$ (which amounts to parameterizing the curve from 1 to 0 instead of from 0 to 1) then we obtain

$$\oint_{-\gamma} g(z)\,dz = \int_1^0 g(t^2 - it) \cdot \gamma'(t)\,dt = \int_1^0 [(t^2 - it)^2 - (t^2 - it)] \cdot (2t - i)\,dt$$

$$= \int_1^0 [t^4 - 2it^3 - 2t^2 + it] \cdot (2t - i)\,dt$$

$$= \int_1^0 2t^5 - 5it^4 - 6t^3 + 4it^2 + t\,dt$$

$$= \left[\frac{2t^6}{6} - \frac{5it^5}{5} - \frac{6t^4}{4} + \frac{4it^3}{3} + \frac{t^2}{2} \right]_1^0$$

$$= -\left(\frac{1}{3} - i - \frac{3}{2} + \frac{4i}{3} + \frac{1}{2} \right)$$

$$= \frac{2}{3} - \frac{i}{3}\,. \hspace{3cm} \square$$

Exercises

1.   In each of the following problems, calculate the complex line integral of the given function f along the given curve γ:

(a) $f(z) = z\bar{z}^2 - \cos z$, $\gamma(t) = \cos 2 + i \sin 2t$, $0 \le t \le \pi/2$

(b) $f(z) = z^2 - \sin \bar{z}$, $\gamma(t) = t + it^2$, $0 \le t \le 1$

(c) $f(z) = z^3 + \frac{z}{z+1}$, $\gamma(t) = e^t + ie^2 t$, $1 \le t \le 2$

(d) $f(z) = e^z - e^{-z}$, $\gamma(t) = t - i \log t$, $1 \le t \le e$

2. Calculate the complex line integral of the holomorphic function $f(z) = z^2$ along the counterclockwise-oriented square of side 2, with sides parallel to the axes, centered at the origin.

3. Calculate the complex line integral of the function $g(z) = 1/z$ along the counterclockwise-oriented square of side 2, with sides parallel to the axes, centered at the origin.

4. Calculate the complex line integral of the holomorphic function $f(z) = z^k$, $k = 0, 1, 2, \ldots$, along the curve $\gamma(t) = \cos t + i \sin t$, $0 \le t \le \pi$. Now calculate the complex line integral of the same function along the curve $\mu(t) = \cos t - i \sin t$, $0 \le t \le \pi$. Verify that, for each fixed k, the two answers are the same.

5. Verify that the conclusion of the last exercise is *false* if we take $k = -1$.

6. Verify that the conclusion of Exercise 4 is still true if we take $k = -2, -3, -4, \ldots$.

7. Suppose that f is a continuous function with antiderivative F. This means that $\partial F / \partial z = f$ on the domain of definition. Let γ be a continuously differentiable, closed curve in the domain of f. Show that

$$\oint_\gamma f(z)\, dz = 0.$$

8. If f is a function and γ is a curve and $\oint_\gamma f(z)\, dz = 0$ then does it follow that $\oint f^2(z)\, dz = 0$?

4.2   The Cauchy Integral Theorem and Formula

4.2.1   The Cauchy Integral Theorem, Basic Form

If f is a holomorphic function on an open disc W in the complex plane, and if $\gamma : [a, b] \to W$ is a C^1 curve in U with $\gamma(a) = \gamma(b)$, then

$$\oint_\gamma f(z)\, dz = 0\,. \tag{4.18}$$

This is the *Cauchy integral theorem*. It is central and fundamental to the theory of complex functions. All of the principal results about holomorphic functions stem from this simple integral formula. We shall spend a good deal of our time in this text studying the Cauchy theorem and its consequences.

We now indicate a reason for this result. In fact it turns out that the Cauchy integral theorem, properly construed, is little more than a restatement of Green's theorem from calculus. Recall (see [THO], [BLK]) that Green's theorem says that if u, v are continuously differentiable on a bounded region U in the plane having C^2 boundary, then

$$\int_{\partial U} u\, dx + v\, dy = \iint_U \left(\frac{\partial v}{\partial x} - \frac{\partial u}{\partial y} \right) dx dy\,. \tag{4.19}$$

In the reasoning that we are about to present, we shall for simplicity assume that the curve γ is simple. That is, γ does not cross itself, so it surrounds a region V. See Figure 4.4. Thus $\gamma = \partial V$. Let us write

$$\oint_\gamma f\, dz = \oint_\gamma (u + iv)\, [dx + idy] = \left(\oint_\gamma u\, dx - v\, dy \right) + i \left(\oint_\gamma v\, dx + u\, dy \right)\,.$$

Each of these integrals is clearly a candidate for application of Green's theorem (4.19). Thus

$$\oint_\gamma f\, dz = \oint_{\partial V} f\, dz = \iint_V \left(\frac{\partial(-v)}{\partial x} - \frac{\partial u}{\partial y} \right) dx dy + i \iint_V \left(\frac{\partial u}{\partial x} - \frac{\partial v}{\partial y} \right) dx dy\,.$$

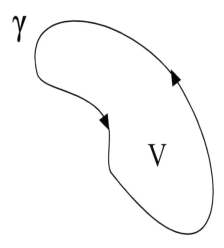

Figure 4.4: The curve γ surrounds the region V.

But, according to the Cauchy–Riemann equations, each of the integrands vanishes. We learn then that

$$\oint_\gamma f \, dz = 0 \, .$$

That is Cauchy's theorem.

An important converse of Cauchy's theorem is called *Morera's theorem*:

Let f be a continuous function on a connected open set $U \subseteq \mathbb{C}$. If

$$\oint_\gamma f(z) \, dz = 0 \qquad\qquad (4.20)$$

for every simple, closed curve γ in U, then f is holomorphic on U.

In the statement of Morera's theorem, the phrase "every simple, closed curve" may be replaced by "every triangle" or "every square" or "every circle." The verification of Morera's theorem also uses Green's theorem. For the very same calculation shows that if

$$\oint_\gamma f(z) \, dz = 0$$

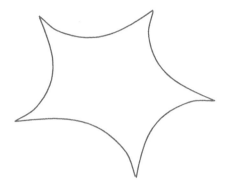

Figure 4.5: A piecewise C^k curve.

for every simple, closed curve γ, then

$$\iint\limits_{U} \left(\frac{\partial(-v)}{\partial x} - \frac{\partial u}{\partial y} \right) + i \left(\frac{\partial u}{\partial x} - \frac{\partial v}{\partial y} \right) \, dx dy = 0$$

for the region U that γ surrounds. This is true for every possible region U! It follows that the integrand must be identically zero. But this simply says that f satisfies the Cauchy–Riemann equations. So it is holomorphic.

4.2.2   More General Forms of the Cauchy Theorem

Now we present the very useful general statements of the Cauchy integral theorem and formula. First we need a piece of terminology. A curve $\gamma : [a, b] \to \mathbb{C}$ is said to be *piecewise* C^k if

$$[a, b] = [a_0, a_1] \cup [a_1, a_2] \cup \cdots \cup [a_{m-1}, a_m] \qquad (4.21)$$

with $a = a_0 < a_1 < \cdots < a_m = b$ and the curve $\gamma|_{[a_{j-1}, a_j]}$ is C^k for $1 \leq j \leq m$. In other words, γ is piecewise C^k if it consists of finitely many C^k curves chained end to end. See Figure 4.5.

Cauchy Integral Theorem: Let $f : U \to \mathbb{C}$ be holomorphic with $U \subseteq \mathbb{C}$ an open set. Then

$$\oint_{\gamma} f(z) \, dz = 0 \qquad (4.22)$$

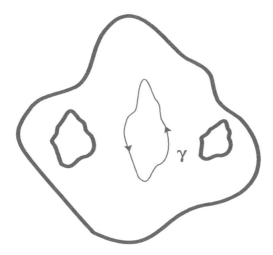

Figure 4.6: A curve γ on which the Cauchy integral theorem is valid.

for each piecewise C^1 closed curve γ in U that can be deformed in U through closed curves to a point in U—see Figure 4.6. We call such a curve *homotopic to 0*. From the topological point of view, such a curve is trivial.

EXAMPLE 5 Let U be the region consisting of the disc $\{z \in \mathbb{C} : |z| < 2\}$ with the closed disc $\{z \in \mathbb{C} : |z-i| < 1/3\}$ removed. Let $\gamma : [0,1] \to U$ be the curve $\gamma(t) = \cos t + [i/10] \sin t$. See Figure 4.7. If f is any holomorphic function on U then

$$\oint_\gamma f(z)\, dz = 0\,.$$

Perhaps more interesting is the following fact. Let P, Q be points of U. Let $\gamma : [0,1] \to U$ be a curve that begins at P and ends at Q. Let $\mu : [0,1] \to U$ be some other curve that begins at P and ends at Q. The requirement that we impose on these curves is that they do not surround any holes in U—in other words, the curve formed with γ followed by (the reverse of) μ is homotopic to 0. Refer to Figure 4.8. If f is any holomorphic function on U then

$$\oint_\gamma f(z)\, dz = \oint_\mu f(z)\, dz\,.$$

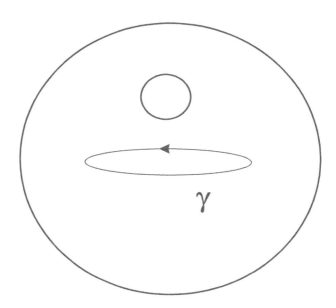

Figure 4.7: A curve γ on which the generalized Cauchy integral theorem is valid.

The reason is that the curve τ that consists of γ *followed by* the reverse of μ is a closed curve in U. It is homotopic to 0. Thus the Cauchy integral theorem applies and

$$\oint_\tau f(z)\,dz = 0\,.$$

Writing this out gives

$$\oint_\gamma f(z)\,dz - \oint_\mu f(z)\,dz = 0\,.$$

That is our claim. $\square$

4.2.3   Deformability of Curves

A central fact about the complex line integral is the deformability of curves. Let $\gamma : [a, b] \to U$ be a piecewise C^1, closed curve in a region U of the complex plane. Let f be a holomorphic function on U. The value

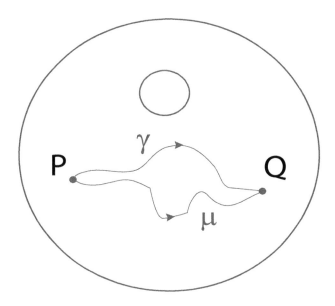

Figure 4.8: Two curves with equal complex line integrals.

of the complex line integral

$$\oint_\gamma f(z)\, dz \tag{4.23}$$

does not change if the curve γ is smoothly deformed within the region U. Note that, in order for this statement to be valid, the curve γ must remain inside the region of holomorphicity U of f while it is being deformed, and it must remain a closed curve while it is being deformed. Figure 4.9 shows curves γ_1, γ_2 that *can* be deformed to one another, and a curve γ_3 that can be deformed to neither of the first two (because of the hole inside γ_3).

The reasoning behind the deformability principle is simplicity itself. Examine Figure 4.10. It shows a solid curve γ and a dashed curve $\widetilde{\gamma}$. The latter should be thought of as a deformation of the former. Now let us examine the *difference* of the integrals over the two curves—see Figure 4.11. We see that this difference is in fact the integral of the holomorphic function f over a closed curve that *can be continuously deformed to a point*. Of course, by the Cauchy integral theorem, that integral is equal to 0. Thus the difference of the integral over γ and the integral over $\widetilde{\gamma}$ is 0. That is the deformability principle.

Figure 4.9: Deformation of curves.

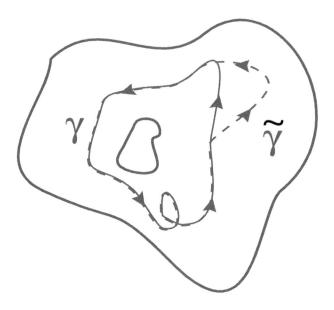

Figure 4.10: Deformation of curves.

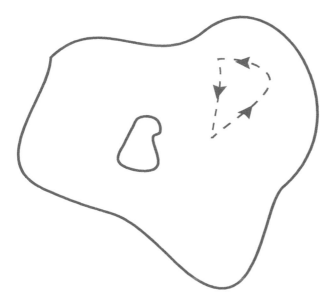

Figure 4.11: The difference of the integrals.

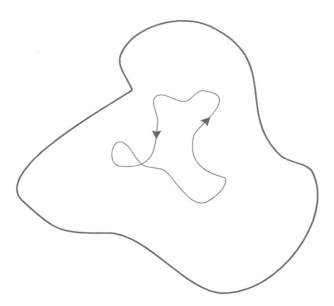

Figure 4.12: A simply connected domain.

Figure 4.13: A domain that is *not* simply connected.

A topological notion that is special to complex analysis is simple connectivity. We say that a domain $U \subseteq \mathbb{C}$ is *simply connected* if any closed curve in U can be continuously deformed to a point. See Figure 4.12. Simple connectivity is a mathematically rigorous condition that corresponds to the intuitive notion that the region U has no holes. Figure 4.13 shows a domain that is *not* simply connected. If U is simply connected, and γ is a closed curve in U, then it follows that γ can be continuously deformed to lie inside a disc in U. It follows that Cauchy's theorem applies to γ. To summarize: on a simply connected region, Cauchy's theorem applies (without any further hypotheses) to any closed curve in U. Likewise, in a simply connected U, Cauchy's integral formula (to be developed below) applies to any simple, closed curve that is oriented counterclockwise and to any point z that is inside that curve.

4.2.4   Cauchy Integral Formula, Basic Form

The Cauchy integral formula is derived from the Cauchy integral theorem. It tells us that we can express the value of a holomorphic function f in terms of a sort of average of its values around a boundary curve. This assertion is really quite profound; it turns out that the formula is key to many of the most important properties of holomorphic functions. We begin with a simple enunciation of Cauchy's idea.

Let $U \subseteq \mathbb{C}$ be a domain and suppose that $\overline{D}(P, r) \subseteq U$. Let $\gamma : [0, 1] \to \mathbb{C}$ be the C^1 parameterization $\gamma(t) = P + r \cos(2\pi t) + ir \sin(2\pi t)$. Then, for each $z \in D(P, r)$,

$$f(z) = \frac{1}{2\pi i} \oint_\gamma \frac{f(\zeta)}{\zeta - z} \, d\zeta. \tag{4.24}$$

Before we indicate the reasoning behind this formula, we impose some simplifications. First, we may as well translate coordinates and assume that $P = 0$. Thus the Cauchy formula becomes

$$f(z) = \frac{1}{2\pi i} \oint_{\partial D(0,r)} \frac{f(\zeta)}{\zeta - z} \, d\zeta.$$

Our strategy is to apply the Cauchy integral *theorem* to the function

$$g(\zeta) = \frac{f(\zeta) - f(z)}{\zeta - z}.$$

In fact it can be checked—using Morera's theorem for example—that g is still holomorphic. Thus we may apply Cauchy's theorem to see that

$$\oint_{\partial D(0,r)} g(\zeta) \, d\zeta = 0$$

or

$$\oint_{\partial D(0,r)} \frac{f(\zeta) - z}{\zeta - z} \, d\zeta = 0.$$

But this just says that

$$\frac{1}{2\pi i} \oint_{\partial D(0,r)} \frac{f(z)}{\zeta - z} \, d\zeta = \frac{1}{2\pi i} \oint_{\partial D(0,r)} \frac{f(\zeta)}{\zeta - z} \, d\zeta. \tag{4.25}$$

It remains to examine the left-hand side.

Now

$$\frac{1}{2\pi i} \oint_{\partial D(0,r)} \frac{f(z)}{\zeta - z} \, d\zeta = \frac{f(z)}{2\pi i} \oint_{\partial D(0,r)} \frac{1}{\zeta - z} \, d\zeta \tag{4.26}$$

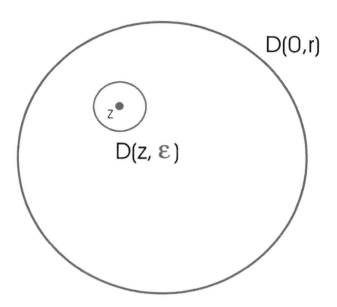

Figure 4.14: Deformability of curves.

and we must evaluate the integral. By deformation of the curve of integration (see Figure 4.14), we have

$$\oint_{\partial D(0,r)} \frac{1}{\zeta - z} \, d\zeta = \oint_{\partial D(z,\epsilon)} \frac{1}{\zeta - z} \, d\zeta = \oint_{\partial D(0,\epsilon)} \frac{1}{\zeta} \, d\zeta . \qquad (4.27)$$

Here $\epsilon > 0$ is a number chosen to be so small that the little disc fits inside the larger disc. Introducing the parameterization $t \mapsto \epsilon e^{it}$, $0 \leq t \leq 2\pi$ for the curve, we find that our integral is

$$\int_0^{2\pi} \frac{1}{e^{it}} \, i e^{it} \, dt = \int_0^{2\pi} i \, dt = 2\pi i .$$

Putting this information together with (4.25) and (4.26), we find that

$$f(z) = \frac{1}{2\pi i} \oint_{\partial D(0,r)} \frac{f(\zeta)}{\zeta - z} \, d\zeta .$$

That is the Cauchy integral formula for a disc.

4.2.5   More General Versions of the Cauchy Formula

A more general version of the Cauchy formula—the one that is typically used in practice—is this:

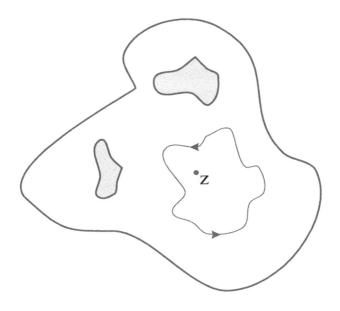

Figure 4.15: Deformability of curves—specifically, circles.

THEOREM 1 *Let $U \subseteq \mathbb{C}$ be a domain. Let $\gamma : [0,1] \to U$ be a simple closed curve that can be continuously deformed to a point inside U. See Figure 4.15. If f is holomorphic on U and z lies in the region interior to γ, then*

$$f(z) = \frac{1}{2\pi i} \oint_\gamma \frac{f(\zeta)}{\zeta - z}\, d\zeta \,.$$

The verification is nearly identical to the one that we have presented above in the special case. We omit the details.

EXAMPLE 6 Let $U = \{z = x + iy \in \mathbb{C} : -2 < x < 2, 0 < y < 3\} \setminus \overline{D}(-1 + 2i, 1/10)$. Let $\gamma(t) = \cos t + i(3/2 + \sin t)$. Then the curve γ lies in U. The curve γ can certainly be deformed to a point inside U. Thus if f is any holomorphic function on U then, for z inside the curve (see Figure 4.16),

$$f(z) = \frac{1}{2\pi i} \oint_\gamma \frac{f(\zeta)}{\zeta - z}\, d\zeta \,. \qquad \square$$

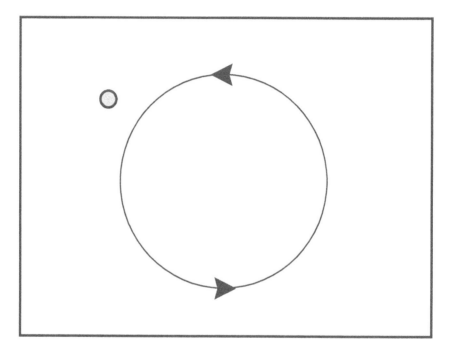

Figure 4.16: Illustration of the Cauchy integral formula.

Exercises

1. Let $f(z) = z^2 - z$ and $\gamma(t) = (\cos t, \sin t)$, $0 \leq t \leq 2\pi$. Confirm the statement of the Cauchy integral theorem for this f and this γ by actually calculating the appropriate complex line integral for z inside the curve.

2. The points $\pm\frac{1}{\sqrt{2}} \pm i\frac{1}{\sqrt{2}}$ lie on the unit circle. Let $\eta(t)$ be the square path with counterclockwise orientation for which they are the vertices. Verify the conclusion of the Cauchy integral theorem for *this* path η and the function $f(z) = z^2 - z$. Compare with Exercise 1.

3. The Cauchy integral theorem fails for the function $f(z) = \cot z \equiv \cos z / \sin z$ on the annulus $\{z \in \mathbb{C} : 1 < |z| < 2\}$. Calculate the relevant complex line integral and verify that the value of the integral is *not* zero. What hypothesis of the Cauchy integral theorem is lacking?

4. Let u be a harmonic function in a neighborhood of the closed unit disc

$$\overline{D}(0,1) = \{z \in \mathbb{C} : |z| \leq 1\}.$$

For each $P = (p_1, p_2) \in \partial D(0,1)$, let $\nu(P) = \langle p_1, p_2 \rangle$ be the unit outward normal vector. Use Green's theorem to show that

$$\int_{\partial D(0,1)} \frac{\partial}{\partial \nu} u(z) \, ds(z) = 0.$$

[**Hint:** Be sure to note that this is *not* a complex line integral. It is instead a standard calculus integral with respect to arc length ds.]

5. It is a fact (Morera's theorem) that if f is a continuously differentiable function on a domain Ω and if $\oint_\gamma f(z) \, dz = 0$ for every continuously differentiable, closed curve in Ω, then f is holomorphic on Ω. Restrict attention to curves that bound closed discs that lie in Ω. Apply Green's theorem to the hypothesis that we have formulated. Conclude that the two-dimensional integral of $\partial f / \partial \overline{z}$ is 0 on any disc in Ω. What does this tell you about $\partial f / \partial \overline{z}$?

6. Let f be holomorphic on a domain Ω and let P, Q be points of Ω. Let γ_1 and γ_2 be continuously differentiable curves in Ω that each begin at P and end at Q. What conditions on γ_1, γ_2, and Ω will guarantee that $\oint_{\gamma_1} f(z) \, dz = \oint_{\gamma_2} f(z) \, dz$?

7.   Let Ω be a domain and suppose that γ is a simple, closed curve in Ω that is continuously differentiable. Suppose that $\oint_\gamma f(z)\, dz = 0$ for every holomorphic function f on Ω. What can you conclude about the domain Ω and the curve γ?

8.   Let D be the unit disc and suppose that $\gamma : [0, 1] \to D$ is a curve that circles the origin *twice* in the counterclockwise direction. Let f be holomorphic on D. What can you say about the value of

$$\oint_\gamma \frac{f(\zeta)}{\zeta - 0}\, d\zeta\, ?$$

9.   Suppose that the curve in the last exercise circles the origin twice in the clockwise direction. Then what can you say about the value of the integral

$$\oint_\gamma \frac{f(\zeta)}{\zeta - 0}\, d\zeta\, ?$$

10.   Let the domain D be the unit disc and let g be a *conjugate holomorphic function* on D (i.e., $\bar{g}$ is holomorphic). Then there exists a simple, closed, continuously differentiable curve γ in D such that $\oint_\gamma g(\zeta)\, d\zeta \neq 0$. Show this assertion.

11.   Let $U = \{z \in \mathbb{C} : 1 < |z| < 4\}$. Let $\gamma(t) = 3\cos t + 3i\sin t$. Let $f(z) = 1/z$. Let $P = 2 + i0$. Verify with a direct calculation that

$$f(P) \neq \frac{1}{2\pi i} \oint_\gamma \frac{f(\zeta)}{\zeta - P}\, d\zeta\, .$$

12.   In the preceding exercise, replace f with $g(\zeta) = \zeta^2$. Now verify that

$$g(P) = \frac{1}{2\pi i} \oint_\gamma \frac{g(\zeta)}{\zeta - P}\, d\zeta\, .$$

Explain why the answer to this exercise is different from the answer to the earlier exercise.

13.   Let $U = D(0, 2)$ and let $\gamma(t) = \cos t + i\sin t$. Verify by a direct calculation that, for any $z \in D(0, 1)$,

$$1 = \frac{1}{2\pi i} \oint_\gamma \frac{1}{\zeta - z}\, d\zeta\, .$$

Now derive the same identity immediately using the Cauchy integral formula with the function $f(z) \equiv 1$.

14. Let $U = \{z \in \mathbb{C} : -4 < x < 4, -4 < y < 4\}$. Let $\gamma(t) = \cos t + i \sin t$. Let $\mu(t) = 2\cos t + 3\sin t$. Finally set $f(z) = z^2$. Of course each of the two curves lies in U. Draw a picture. Let $P = 1/2 + i/2$. Calculate

$$\frac{1}{2\pi i} \oint_\gamma \frac{f(\zeta)}{\zeta - P} \, d\zeta$$

and

$$\frac{1}{2\pi i} \oint_\mu \frac{f(\zeta)}{\zeta - P} \, d\zeta$$

The answers that you obtain should be the same. Explain why.

4.3   Variants of the Cauchy Formula

The Cauchy formula is a remarkably flexible tool that can be applied even when the domain U in question is *not* simply connected. Rather than attempting to formulate a general result, we illustrate the ideas here with some examples.

EXAMPLE 7 Let $U = \{z \in \mathbb{C} : 1 < |z| < 4\}$. Let $\gamma_1(t) = 2\cos t + 2i\sin t$ and $\gamma_2(t) = 3\cos t + 3i\sin t$. See Figure 4.17. If f is any holomorphic function on U and if the point z satisfies $2 < |z| < 3$ (again see Figure 4.17) then

$$f(z) = \frac{1}{2\pi i} \oint_{\gamma_2} \frac{f(\zeta)}{\zeta - z} \, d\zeta - \frac{1}{2\pi i} \oint_{\gamma_1} \frac{f(\zeta)}{\zeta - z} \, d\zeta. \qquad (4.28)$$

The beauty of this result is that it can be established with a simple diagram. Refer to Figure 4.18. We see that integration over γ_2 and $-\gamma_1$, as indicated in formula (4.28), is just the same as integrating over a single contour γ^*. And, with a slight deformation, we see that that contour is equivalent—for the purposes of integration—with integration over a contour $\widetilde{\gamma}^*$ that is homotopic to zero. Thus, with a bit of manipulation, we see that the integrations in (4.28) are equivalent to integration of a curve for which we know that the Cauchy formula holds.

That establishes formula (4.28). □

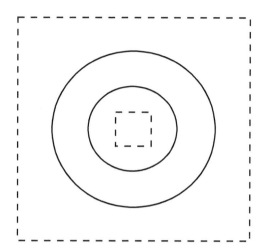

Figure 4.17: A variant of the Cauchy integral formula.

EXAMPLE 8 Consider the region

$$U = D(0,6) \setminus \left[\overline{D}(-3+0i,1) \cup \overline{D}(3+0i,1) \right] .$$

It is depicted in Figure 4.19. We also show in the figure *three* contours of integration: γ_1, γ_2, γ_3. We deliberately do not give formulas for these curves, because we want to stress that the reasoning here is geometric and does *not* depend on formulas.

Now suppose that f is a holomorphic function on U. We want to write a Cauchy integral formula—for the function f and the point z—that will be valid in this situation. It turns out that the correct formula is

$$f(z) = \frac{1}{2\pi i} \oint_{\gamma_1} \frac{f(\zeta)}{\zeta - z} \, d\zeta - \frac{1}{2\pi i} \oint_{\gamma_2} \frac{f(\zeta)}{\zeta - z} \, d\zeta - \frac{1}{2\pi i} \oint_{\gamma_3} \frac{f(\zeta)}{\zeta - z} \, d\zeta .$$

The justification, parallel to that in the last example, is shown in Figure 4.20.   □

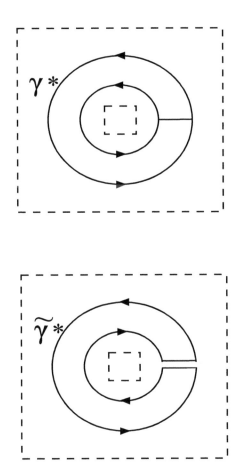

Figure 4.18: Turning two contours into one.

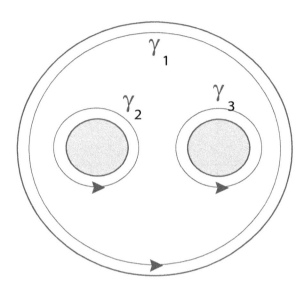

Figure 4.19: A triply connected domain.

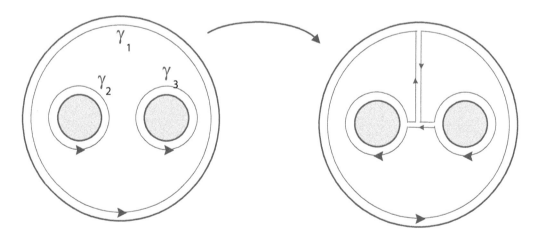

Figure 4.20: Turning three contours into one.

4.4 A Coda on the Limitations of the Cauchy Integral Formula

If f is any continuous function on the boundary of the unit disc $D = D(0,1)$, then the Cauchy integral

$$\frac{1}{2\pi i} \oint_{\partial D} \frac{f(\zeta)}{\zeta - z} \, d\zeta \tag{4.29}$$

defines a holomorphic function $F(z)$ on D (use Morera's theorem, for example, to confirm this assertion). What does the new function F have to do with the original function f? In general, not much.

For example, if $f(\zeta) = \overline{\zeta}$, then $F(z) \equiv 0$ (exercise). In no sense is the original function f any kind of "boundary limit" of the new function F. The question of which functions f are "natural boundary functions" for holomorphic functions F (in the sense that F is a continuous extension of F to the closed disc) is rather subtle. Its answer is well understood, but is best formulated in terms of Fourier series and the so-called Hilbert transform. The complete story is given in [KRA1].

Contrast this situation for holomorphic function with the much more succinct and clean situation for harmonic functions (§13.3).

Exercises

1. Let $U = D(0,4) \setminus \overline{D}(0,1)$. Define $\gamma(t) = 3e^{it}$, $0 \le t < 2\pi$. And define $\mu(t) = 2e^{2it}$, $0 \le t < \pi$. Let f be a holomorphic function on U. Let $2 < |z| < 3$. Explain why

$$\frac{1}{2\pi i} \oint_\gamma \frac{f(\zeta)}{\zeta - z} \, d\zeta = \frac{1}{2\pi i} \oint_\mu \frac{f(\zeta)}{\zeta - z} \, d\zeta \, .$$

2. Let $U = D(0,4)$. Set $\gamma(t) = (2+2i) + e^{it}$ and set $\mu(t) = (-2-2i) - e^{it}$, $0 \le t < 2\pi$. Let f be holomorphic on U and let $z \in D((2-2i), 1)$. Explain why

$$\frac{1}{2\pi i} \oint_\gamma \frac{f(\zeta)}{\zeta - z} \, d\zeta = \frac{1}{2\pi i} \oint_\mu \frac{f(\zeta)}{\zeta - z} \, d\zeta \, .$$

3. Let $U = D(0,4)$. Set $\gamma(t) = (2+2i) + e^{it}$ and set $\mu(t) = (-2-2i) - e^{it}$, $0 \leq t < 2\pi$. Let f be holomorphic on U and nonvanishing and let $z \in D((2+2i),1)$. Explain why

$$\frac{1}{2\pi i} \oint_\gamma \frac{f(\zeta)}{\zeta - z}\, d\zeta \neq \frac{1}{2\pi i} \oint_\mu \frac{f(\zeta)}{\zeta - z}\, d\zeta \,.$$

4. Let $U = D(0,4) \setminus \overline{D}(0,1)$. Define $\gamma(t) = 3e^{it}$, $0 \leq t < 2\pi$. And define $\mu(t) = 2e^{2it}$, $0 \leq t < \pi$. Let f be a holomorphic function on U. Let $1 < |z| < 2$. Explain why

$$\frac{1}{2\pi i} \oint_\gamma \frac{f(\zeta)}{\zeta - z}\, d\zeta = \frac{1}{2\pi i} \oint_\mu \frac{f(\zeta)}{\zeta - z}\, d\zeta \,.$$

5. Let $U = D(0,4) \setminus \overline{D}(0,1)$. Define $\gamma(t) = 3e^{it}$, $0 \leq t < 2\pi$. And define $\mu(t) = 2e^{2it}$, $0 \leq t < 2\pi$. Let f be a holomorphic function on U. Let $2 < |z| < 3$. Explain why

$$\frac{1}{2\pi i} \oint_\gamma \frac{f(\zeta)}{\zeta - z}\, d\zeta \neq \frac{1}{2\pi i} \oint_\mu \frac{f(\zeta)}{\zeta - z}\, d\zeta \,.$$

6. Let k be a nonnegative integer. Explicitly calculate

$$\frac{1}{2\pi i} \oint_{\partial D(0,1)} \overline{\zeta}^k \, d\zeta \,.$$

7. Let k be a negative integer. Explicitly calculate

$$\frac{1}{2\pi i} \oint_{\partial D(0,1)} \overline{\zeta}^k \, d\zeta \,.$$

8. Let k be a negative integer. Explicitly calculate

$$\frac{1}{2\pi i} \oint_{\partial D(0,1)} \zeta^k \, d\zeta \,.$$

9. Explicitly calculate

$$\frac{1}{2\pi i} \oint_{\partial D(0,1)} \frac{\overline{\zeta}}{\zeta} \, d\zeta \,.$$

Chapter 5

Applications of the Cauchy Theory

5.1 The Derivatives of a Holomorphic Function

One of the remarkable features of holomorphic function theory is that we can express the derivative of a holomorphic function in terms of the function itself. Nothing of the sort is true for real functions. One upshot is that we can obtain powerful estimates for the derivatives of holomorphic functions.

We shall explore this phenomenon in the present section.

EXAMPLE 1 On the real line $\mathbb{R}$, let

$$f_k(x) = \sin(kx).$$

Then of course $|f_k(x)| \leq 1$ for all k and all x. Yet $f_k'(x) = k\cos(kx)$ and $|f_k'(0)| = k$. So there is no sense, and no hope, of bounding the derivative of a function by means of the function itself. We will find matters to be quite different for holomorphic functions. $\qquad \square$

5.1.1   A Formula for the Derivative

Let $U \subseteq \mathbb{C}$ be an open set and let f be holomorphic on U. Then $f \in C^\infty(U)$. Moreover, if $\overline{D}(P, r) \subseteq U$ and $z \in D(P, r)$, then

$$\left(\frac{d}{dz}\right)^k f(z) = \frac{k!}{2\pi i} \oint_{|\zeta - P| = r} \frac{f(\zeta)}{(\zeta - z)^{k+1}}\, d\zeta, \quad k = 0, 1, 2, \ldots. \qquad (5.1)$$

The verification of this new formula is direct. For consider the Cauchy formula:

$$f(z) = \frac{1}{2\pi i} \oint_{\partial D(0, r)} \frac{f(\zeta)}{\zeta - z}\, d\zeta.$$

We may differentiate both sides of this equation:

$$\frac{d}{dz} f(z) = \frac{d}{dz} \left[\frac{1}{2\pi i} \oint_{\partial D(0, r)} \frac{f(\zeta)}{\zeta - z}\, d\zeta \right].$$

Now we wish to justify passing the derivative on the right under the integral sign. A justification from first principles may be obtained by examining the Newton quotients for the derivative. Alternatively, one can cite a suitable limit theorem as in [RUD1] or [KRA2]. In any event, we obtain

$$\begin{aligned}
\frac{d}{dz} f(z) &= \frac{1}{2\pi i} \oint_{\partial D(0, r)} \frac{d}{dz} \left[\frac{f(\zeta)}{\zeta - z} \right] d\zeta \\
&= \frac{1}{2\pi i} \oint_{\partial D(0, r)} f(\zeta) \cdot \frac{d}{dz} \left[\frac{1}{\zeta - z} \right] d\zeta \\
&= \frac{1}{2\pi i} \oint_{\partial D(0, r)} f(\zeta) \cdot \frac{1}{(\zeta - z)^2}\, d\zeta.
\end{aligned}$$

This is in fact the special instance of formula (5.1) when $k = 1$. The cases of higher k are obtained through additional differentiations, or by induction.

5.1.2   The Cauchy Estimates

If f is a holomorphic on a region containing the closed disc $\overline{D}(P, r)$ and if $|f| \leq M$ on $\overline{D}(P, r)$, then

$$\left| \frac{\partial^k}{\partial z^k} f(P) \right| \leq \frac{M \cdot k!}{r^k}. \qquad (5.2)$$

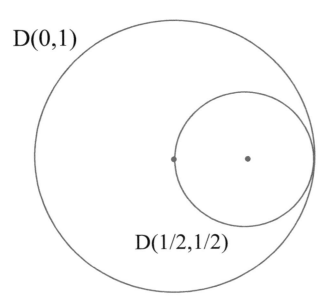

D(0,1)

D(1/2,1/2)

Figure 5.1: The Cauchy estimates.

These are called the *Cauchy estimates.*

In fact this formula is a result of direct estimation from (5.1). For we have

$$\left| \frac{\partial^k}{\partial z^k} f(P) \right| = \left| \frac{k!}{2\pi i} \oint_{|\zeta - P| = r} \frac{f(\zeta)}{(\zeta - z)^{k+1}} \, d\zeta \right| \leq \frac{k!}{2\pi} \cdot \frac{M}{r^{k+1}} \cdot 2\pi r = \frac{Mk!}{r^k} \, .$$

EXAMPLE 2 Let $f(z) = (z^3 + 1)e^{z^2}$ on the unit disc $D(0,1)$. Obviously

$$|f(z)| \leq 2 \cdot |e^{z^2}| = e^{x^2 - y^2} \leq e \quad \text{for all } z \in D(0,1) \, .$$

We may then conclude, by the Cauchy estimates applied to f on $D(1/2, 1/2) \subseteq D(0,1)$ (see Figure 5.1), that

$$|f'(1/2)| \leq \frac{e \cdot 1!}{1/2} = e$$

and

$$|f''(1/2)| \leq \frac{e \cdot 2!}{(1/2)^2} = 8e \, .$$

Of course one may perform the tedious calculation of these derivatives and determine that $f'(1/2) \approx 1.1235$ and $f''(1/2) \approx 6.2596$. But

Cauchy's estimates allow us to estimate the derivatives by way of soft analysis. $\square$

5.1.3   Entire Functions and Liouville's Theorem

A function f is said to be *entire* if it is defined and holomorphic on all of $\mathbb{C}$, i.e., $f : \mathbb{C} \to \mathbb{C}$ is holomorphic. For instance, any holomorphic polynomial is entire, e^z is entire, and $\sin z, \cos z$ are entire. The function $f(z) = 1/z$ is not entire because it is undefined at $z = 0$. [In a sense that we shall make precise later (§6.1), this last function has a "singularity" at 0.] The question we wish to consider is: "Which entire functions are bounded?" This question has a very elegant and complete answer as follows:

THEOREM 1 (Liouville's Theorem) *A bounded entire function is* *constant.*

To understand this result, let f be entire and assume that $|f(z)| \le M$ for all $z \in \mathbb{C}$. Fix a $P \in \mathbb{C}$ and let $r > 0$. We apply the Cauchy estimate (5.2) for $k = 1$ on $\overline{D}(P, r)$. So

$$\left| \frac{\partial}{\partial z} f(P) \right| \le \frac{M \cdot 1!}{r}.$$

Since this inequality is true for every $r > 0$, we conclude that

$$\frac{\partial f}{\partial z}(P) = 0.$$

Since P was arbitrary, we conclude that

$$\frac{\partial f}{\partial z} \equiv 0.$$

Of course we also know, since f is holomorphic, that

$$\frac{\partial f}{\partial \overline{z}} \equiv 0.$$

It follows from linear algebra then that

$$\frac{\partial f}{\partial x} \equiv 0 \qquad \text{and} \qquad \frac{\partial f}{\partial y} \equiv 0 \,.$$

Therefore f is constant.

The reasoning that establishes Liouville's theorem can also be used to demonstrate this more general fact: If $f : \mathbb{C} \to \mathbb{C}$ is an entire function and if for some real number C and some positive integer k, it holds that

$$|f(z)| \leq C \cdot (1 + |z|)^k \tag{5.3}$$

for all z, then f is a polynomial in z of degree at most k. We leave the details for the interested reader.

5.1.4   The Fundamental Theorem of Algebra

One of the most elegant applications of Liouville's Theorem is a verification of what is known as the Fundamental Theorem of Algebra (see also §§1.2.4, 5.1.4, 9.3.3):

> **The Fundamental Theorem of Algebra:** Let $p(z)$ be a non-constant (holomorphic) polynomial. Then p has a root. That is, there exists an $\alpha \in \mathbb{C}$ such that $p(\alpha) = 0$.

To understand this result, suppose not. Then $g(z) = 1/p(z)$ is entire. Also, when $|z| \to \infty$, then $|p(z)| \to +\infty$. Thus $1/|p(z)| \to 0$ as $|z| \to \infty$; hence g is bounded. By Liouville's Theorem, g is constant, hence p is constant. Contradiction.

If, in the theorem, p has degree $k \geq 1$, then let α_1 denote the root provided by the Fundamental Theorem. By the Euclidean algorithm (see [HUN]), we may divide $z - \alpha_1$ into p to obtain

$$p(z) = (z - \alpha_1) \cdot p_1(z) + r_1(z). \tag{5.4}$$

Here p_1 is a polynomial of degree $k - 1$ and r_1 is the remainder term of degree 0 (i.e., less than 1). Substitutingr α_1 into this last equation gives $0 = 0 + r_1$, hence we see that $r_1 = 0$. Thus the Euclidean algorithm has taught us that

$$p(z) = (z - \alpha_1) \cdot p_1(z).$$

If $k - 1 \geq 1$, then, by the Fundamental Theorem, p_1 has a root α_2 . Thus p_1 is divisible by $(z - \alpha_2)$ and we have

$$p(z) = (z - \alpha_1) \cdot (z - \alpha_2) \cdot p_2(z) \tag{5.5}$$

for some polynomial $p_2(z)$ of degree $k - 2$. This process can be continued until we arrive at a polynomial p_k of degree 0; that is, p_k is constant. We have derived the following fact: If $p(z)$ is a holomorphic polynomial of degree k, then there are k complex numbers $\alpha_1, \ldots \alpha_k$ (not necessarily distinct) and a non-zero constant C such that

$$p(z) = C \cdot (z - \alpha_1) \cdots (z - \alpha_k). \tag{5.6}$$

If some of the roots of p coincide, then we say that p has *multiple roots*. To be specific, if m of the values $\alpha_{j_1}, \ldots, \alpha_{j_m}$ are equal to some complex number α, then we say that p has a root of order m at α (or that p has a root α of *multiplicity m*). An example will make the idea clear: Let

$$p(z) = (z - 5)^3 \cdot (z + 2)^8 \cdot (z - 7) \cdot (z + 6). \tag{5.7}$$

This polynomial p has degree 13. Then we say that p has a root of order 3 at 5, a root of order 8 at -2, and it has roots of order 1 at 7 and at -6. We also say that p has *simple roots* at 7 and -6.

5.1.5   Sequences of Holomorphic Functions and Their Derivatives

A sequence of functions g_j defined on a common domain E is said to *converge uniformly* to a limit function g if, for each $\epsilon > 0$, there is a number $N > 0$ such that, for all $j > N$, it holds that $|g_j(x) - g(x)| < \epsilon$ for every $x \in E$. The key point is that the degree of closeness of $g_j(x)$ to $g(x)$ is independent of $x \in E$.

Let $f_j : U \to \mathbb{C}$, $j = 1, 2, 3 \ldots$, be a sequence of holomorphic functions on an open set $U \subseteq \mathbb{C}$. Suppose that there is a function $f : U \to \mathbb{C}$ such that, for each compact subset E (a compact set is one that is closed and bounded—see Figure 5.2) of U, the restricted sequence $f_j|_E$ converges uniformly to $f|_E$. Then f is holomorphic on U. [In particular, $f \in C^\infty(U)$.]

One may see this last assertion by examining the Cauchy integral formula:

$$f_j(z) = \frac{1}{2\pi i} \oint \frac{f_j(\zeta)}{\zeta - z} \, d\zeta.$$

Now we may let $j \to \infty$, and invoke the uniform convergence to pass the limit under the integral sign on the right. The result is

$$
\begin{aligned}
\lim_{j\to\infty} f_j(z) &= \lim_{j\to\infty} \frac{1}{2\pi i} \oint \frac{f_j(\zeta)}{\zeta - z} d\zeta \\
&= \frac{1}{2\pi i} \oint \lim_{j\to\infty} \frac{f_j(\zeta)}{\zeta - z} d\zeta \\
&= \frac{1}{2\pi i} \oint \frac{\lim_{j\to\infty} f_j(\zeta)}{\zeta - z} d\zeta
\end{aligned}
$$

or

$$
f(z) = \frac{1}{2\pi i} \oint \frac{f(\zeta)}{\zeta - z} d\zeta .
$$

The right-hand side is plainly a holomorphic function of z (simply differentiate under the integral sign, or apply Morera's theorem). Thus f is holomorphic.

If f_j, f, U are as in the preceding paragraph, then, for any $k \in \{0, 1, 2, \dots\}$, we have

$$
\left(\frac{\partial}{\partial z}\right)^k f_j(z) \to \left(\frac{\partial}{\partial z}\right)^k f(z) \tag{5.8}
$$

uniformly on compact sets. This again follows from an examination of the Cauchy integral formula. We omit the details.

5.1.6 The Power Series Representation of a Holomorphic Function

The ideas being considered in this section can be used to develop our understanding of power series. A *power series*

$$
\sum_{j=0}^{\infty} a_j(z - P)^j \tag{5.9}
$$

is defined to be the limit of its *partial sums*

$$
S_N(z) = \sum_{j=0}^{N} a_j(z - P)^j. \tag{5.10}
$$

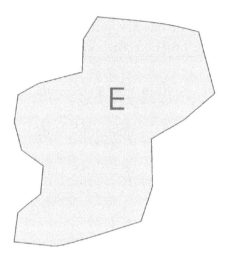

Figure 5.2: A compact set is closed and bounded.

We say that the partial sums *converge* to the sum of the entire series.

Any given power series has a *disc of convergence*. More precisely, let

$$r = \frac{1}{\limsup_{j \to \infty} |a_j|^{1/j}}. \tag{5.11}$$

The power series (5.10) will then certainly converge on the disc $D(P,r)$; the convergence will be absolute and uniform on any disc $\overline{D}(P,r')$ with $r' < r$.

For clarity, we should point out that in many examples the sequence $|a_j|^{1/j}$ actually converges as $j \to \infty$. Then we may take r to be equal to $1/\lim_{j\to\infty} |a_j|^{1/j}$. The reader should be aware, however, that in case the sequence $\{|a_j|^{1/j}\}$ does not converge, then one must use the more formal definition (5.11) of r. See [KRA2], [RUD1].

Of course the partial sums, being polynomials, are holomorphic on *any* disc $D(P,r)$. If the disc of convergence of the power series is $D(P,r)$, then let f denote the function to which the power series converges. Then, for any $0 < r' < r$, we have that

$$S_N(z) \to f(z), \tag{5.12}$$

uniformly on $\overline{D}(P,r')$. We can conclude immediately that $f(z)$ is holomorphic on $D(P,r')$ for any $r' < r$ hence on $D(P,r)$. Moreover, we know that

$$\left(\frac{\partial}{\partial z}\right)^k S_N(z) \to \left(\frac{\partial}{\partial z}\right)^k f(z). \tag{5.13}$$

This shows that a differentiated power series has a disc of convergence at least as large as the disc of convergence (with the same center) of the original series, and that the differentiated power series converges on that disc to the derivative of the sum of the original series. In fact, the differentiated series has exactly the same radius of convergence as the original series.

The most important fact about power series for complex function theory is this: If f is a holomorphic function on a domain $U \subseteq \mathbb{C}$, if $P \in U$, and if the disc $D(P, r)$ lies in U, then f may be represented as a convergent power series on $D(P, r)$. Explicitly, we have

$$f(z) = \sum_{j=0}^{\infty} a_j (z - P)^j. \tag{5.14}$$

But how do we actually calculate the power series expansion of a given holomorphic function? If we suppose in advance that f has a convergent power series expansion on the disc $D(P, r)$, then we may write

$$f(z) = a_0 + a_1(z - P) + a_2(z - P)^2 + a_3(z - P)^3 + \cdots. \tag{5.15}$$

Now let us evaluate both sides at $z = P$. We see immediately that $f(P) = a_0$.

Next, differentiate both sides of (5.15). The result is

$$f'(z) = a_1 + 2a_2(z - P) + 3a_3(z - P)^2 + \cdots.$$

Again, evaluate both sides at $z = P$. The result is $f'(P) = a_1$.

We may differentiate one more time and evaluate at $z = P$ to learn that $f''(P) = 2a_2$. Continuing in this manner, we discover that $f^{(k)}(P) = k! a_k$, where the superscript (k) denotes k derivatives.

We have discovered a convenient and elegant formula for the power series coefficients:

$$a_k = \frac{f^{(k)}(P)}{k!}. \tag{5.16}$$

EXAMPLE 3 Let us determine the power series for $f(z) = z \sin z$ expanded about the point $P = \pi$. We begin by calculating

$$
\begin{aligned}
f'(z) &= \sin z + z \cos z \\
f''(z) &= 2 \cos z - z \sin z \\
f'''(z) &= -3 \sin z - z \cos z \\
f^{(iv)} &= -4 \cos z + z \sin z
\end{aligned}
$$

and, in general,

$$f^{(2\ell+1)}(z) = (-1)^\ell (2\ell + 1) \sin z + (-1)^\ell z \cos z$$

and

$$f^{(2\ell)}(z) = (-1)^{\ell+1}(2\ell) \cos z + (-1)^\ell z \sin z \,.$$

Evaluating at π, and using formula (5.16), we find that

$$
\begin{aligned}
a_0 &= 0 \\
a_1 &= -\pi \\
a_2 &= -1 \\
a_3 &= \frac{\pi}{3!} \\
a_4 &= \frac{1}{3!} \\
a_5 &= -\frac{\pi}{5!} \\
a_6 &= -\frac{1}{5!} \,,
\end{aligned}
$$

and, in general,

$$a_{2\ell} = \frac{(-1)^\ell}{(2\ell - 1)!}$$

and

$$a_{2\ell+1} = (-1)^{\ell+1} \frac{\pi}{2\ell + 1} \,.$$

In conclusion, the power series expansion for $f(z) = z \sin z$, expanded about the point $P = \pi$, is

$$f(z) = -\pi(z - \pi) - (z - \pi)^2 + \frac{\pi}{3!} \cdot (z - \pi)^3 + \frac{1}{3!} \cdot (z - \pi)^4$$

$$- \frac{\pi}{5!} \cdot (z - \pi)^5 - \frac{1}{5!} \cdot (z - 5)^6 + - \cdots$$

$$= \pi \sum_{\ell=0}^{\infty} (-1)^{\ell+1} \frac{(z - \pi)^{2\ell+1}}{(2\ell + 1)!} + \sum_{\ell=1}^{\infty} (-1)^{\ell} \frac{(z - \pi)^{2\ell}}{(2\ell - 1)!}. \qquad \square$$

In summary, we have an explicit way of calculating the power series expansion of any holomorphic function f about a point P of its domain, and we have an *a priori* knowledge of the disc on which the power series representation will converge.

5.1.7 Table of Elementary Power Series

The table on the following page presents a summary of elementary power series expansions.

Table of Elementary Power Series

Function	Power Series abt. 0	Disc of Convergence
$\dfrac{1}{1 - z}$	$\displaystyle\sum_{n=0}^{\infty} z^n$	$\{z : \lvert z \rvert < 1\}$
$\dfrac{1}{(1 - z)^2}$	$\displaystyle\sum_{n=1}^{\infty} n z^{n-1}$	$\{z : \lvert z \rvert < 1\}$
$\cos z$	$\displaystyle\sum_{n=0}^{\infty} (-1)^n \frac{z^{2n}}{(2n)!}$	all z
$\sin z$	$\displaystyle\sum_{n=0}^{\infty} (-1)^n \frac{z^{2n+1}}{(2n + 1)!}$	all z
e^z	$\displaystyle\sum_{n=0}^{\infty} \frac{z^n}{n!}$	all z
$\log(z + 1)$	$\displaystyle\sum_{n=0}^{\infty} \frac{(-1)^n}{n + 1} z^{n+1}$	$\{z : \lvert z \rvert < 1\}$
$(z + 1)^{\beta}$	$\displaystyle\sum_{n=0}^{\infty} \binom{\beta}{n} z^n$	$\{z : \lvert z \rvert < 1\}$

Exercises

1.   Suppose that f is an entire function, k is a positive integer, and

$$|f(z)| \le C(1 + |z|^k)$$

for all $z \in \mathbb{C}$. Show that f must be a polynomial of degree at most k.

2.   Suppose that f is an entire function, p is a polynomial, and f/p is bounded. What can you conclude about f?

3.   Let $0 < m < k$ be integers. Give an example of a polynomial of degree k that has just m distinct roots.

4.   Suppose that the polynomial p has a double root at the complex value z_0. Show that $p(z_0) = 0$ and $p'(z_0) = 0$.

5.   Suppose that the polynomial p has a simple zero at z_0 and let γ be a simple closed, continuously differentiable curve that encircles z_0 (oriented in the counterclockwise direction). What can you say about the value of

$$\frac{1}{2\pi i} \oint_\gamma \frac{p'(\zeta)}{p(\zeta)} \, d\zeta \, ?$$

6.   Let $\Omega \subseteq \mathbb{C}$ be a domain and let $\{f_j\}$ be holomorphic functions on Ω. Assume that the sequence $\{f_j\}$ converges uniformly on Ω. Show that, if K is any closed, bounded set in Ω and m is a positive integer, then the sequence $f_j^{(m)}$ (the sequence of mth derivatives) converge will uniformly on K.

7.   Demonstrate a version of the Cauchy estimates for harmonic functions.

8.   For each k, M, r, give an example to show that the Cauchy estimates are sharp.

9.   Demonstrate this sharpening of Liouville's theorem: *If f is an entire function and $|f(z)| \le C|z|^{1/2} + D$ for all z and for some constants C, D then f is constant.* How large can you make the exponent $1/2$ and still draw the same conclusion?

10. Suppose that $p(z)$ is a polynomial of degree k with leading coefficient 1. Assume that all the zeros of p lie in unit disc. Show that, for z large, $|p(z)| \geq 9|z|^k/10$.

11. Let f be a holomorphic function defined on some open region $U \subseteq \mathbb{C}$. Fix a point $P \in U$. Show that the power series expansion of f about P will converge absolutely and uniformly on any disc $\overline{D}(P,r)$ with $r < \text{dist}(P, \partial U)$.

12. Let $0 \leq r \leq \infty$. Give an example of a complex power series, centered at $P \in \mathbb{C}$, with radius of convergence precisely r.

13. We know that
$$\frac{1}{1-z} = 1 + z + z^2 + z^3 + \cdots .$$

Use this model, together with differentiation of series, to find the power series expansion about 0 for

$$\frac{1}{(1-w^2)^2} .$$

14. Use the idea of the last exercise to find the power series expansion about 0 of the function
$$\frac{(1-z^2)^2}{(1+z^2)^2} .$$

5.2   The Zeros of a Holomorphic Function

5.2.1   The Zero Set of a Holomorphic Function

Let f be a holomorphic function. If f is not identically zero, then it turns out that f cannot vanish at too many points. This once again bears out the dictum that holomorphic functions are a lot like polynomials. To give this notion a precise formulation, we need to recall the topological notion of connectedness (§§1.2.2). An open set $W \subseteq \mathbb{C}$ is *connected* if it is not possible to find two disjoint, non-empty open sets U, V in $\mathbb{C}$ such that

$$W = (U \cap W) \cup (V \cap W). \tag{5.17}$$

[In the special context of open sets in the plane, it turns out that connectedness is equivalent to the condition that any two points of W may be connected by a curve that lies entirely in W—see the discussion in §§1.2.2 on path-connectedness.] Now we have:

Discreteness of the Zeros of a Holomorphic Function

> Let $U \subseteq \mathbb{C}$ be a connected (§§1.2.2) open set and let $f : U \to \mathbb{C}$ be holomorphic. Let the zero set of f be $\mathcal{Z} = \{z \in U : f(z) = 0\}$. If there are a $z_0 \in \mathcal{Z}$ and $\{z_j\}_{j=1}^{\infty} \subseteq \mathcal{Z} \setminus \{z_0\}$ such that $z_j \to z_0$, then $f \equiv 0$ on U.

Let us formulate the result in topological terms. We recall that a point z_0 is said to be an *accumulation point* of a set $\mathcal{Z}$ if there is a sequence $\{z_j\} \subseteq \mathcal{Z} \setminus \{z_0\}$ with $\lim_{j\to\infty} z_j = z_0$. Then the theorem is equivalent to the statement: If $f : U \to \mathbb{C}$ is a holomorphic function on a connected (§§1.2.2) open set U and if $\mathcal{Z} = \{z \in U : f(z) = 0\}$ has an accumulation point *in* U, then $f \equiv 0$.

5.2.2   Discrete Sets and Zero Sets

There is still more terminology concerning the zero set of a holomorphic function in §§5.2.1. A set S is said to be *discrete* if for each $s \in S$ there is an $\epsilon > 0$ such that $D(s, \epsilon) \cap S = \{s\}$. See Figure 5.3. People also say, in a slight abuse of language, that a discrete set has points that are "isolated" or that S contains only "isolated points." The result in §§5.2.1 thus asserts that, if f is a non-constant holomorphic function on a connected open set, then its zero set is discrete or, less formally, the zeros of f are isolated.

EXAMPLE 4 It is important to realize that the result in §§5.2.1 does *not* rule out the possibility that the zero set of f can have accumulation points in $\mathbb{C} \setminus U$; in particular, a non-constant holomorphic function on an open set U can indeed have zeros accumulating at a point of ∂U. Consider, for instance, the function $f(z) = \sin(1/[1-z])$ on the unit disc. The zeros of this f include $\{1 - 1/[j\pi]\}$, and these accumulate at the boundary point 1. See Figure 5.4.   □

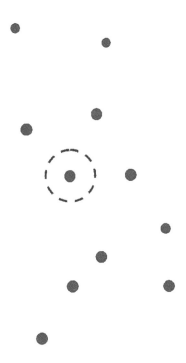

Figure 5.3: A discrete set.

EXAMPLE 5 The function $g(z) = \sin z$ has zeros at $z = k\pi$. Since the domain of g is the entire plane, these infinitely many zeros have no accumulation point so there is no contradiction in that g is not identically zero.

By contrast, the domain $U = \{z = x+iy \in \mathbb{C} : -1 < x < 1, -1 < y < 1\}$ is bounded. If f is holomorphic on U then a holomorphic f can only have finitely many zeros in any compact subset of U. If a holomorphic g has infinitely many zeros, then those zeros can only accumulate at a boundary point. Examples are

$$f(z) = \left(z - \frac{1}{2}\right)^2 \cdot \left(z + \frac{i}{2}\right)^3,$$

with zeros at $1/2$ and $-i/2$, and

$$g(z) = \cos\left(\frac{i}{i-z}\right).$$

Notice that the zeros of g are at $z_k = i\frac{(2k+1)\pi-2}{(2k+1)\pi}$. There are infinitely

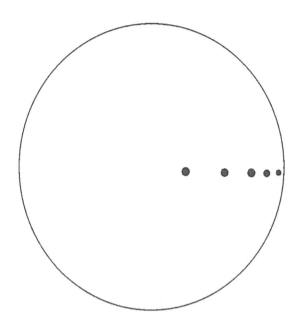

Figure 5.4: A zero set with a boundary accumulation point.

many of these zeros, and they accumulate *only* at i.   □

5.2.3   Uniqueness of Analytic Continuation

A consequence of the preceding basic fact (§§5.2.1) about the zeros of a holomorphic function is this: Let $U \subseteq \mathbb{C}$ be a connected open set and $D(P,r) \subseteq U$. If f is holomorphic on U and $f\big|_{D(P,r)} \equiv 0$, then we may conclude that $f \equiv 0$ on U. This is so because the disc $D(P,r)$ certainly contains an interior accumulation point (merely take $z_j = P + r/j$ and $z_j \to z_0 = P$) hence f must be identically equal to 0.

Here are some further corollaries:

1.   Let $U \subseteq \mathbb{C}$ be a connected open set.   Let f, g be holomorphic on U. If $\{z \in U : f(z) = g(z)\}$ has an accumulation point in U, then $f \equiv g$. Just apply our uniqueness result to the difference function $h(z) = f(z) - g(z)$.

2. Let $U \subseteq \mathbb{C}$ be a connected open set and let f, g be holomorphic on U. If $f \cdot g \equiv 0$ on U, then either $f \equiv 0$ on U or $g \equiv 0$ on U. To see this, we notice that if neither f nor g is identically 0 then there is either a point p at which $f(p) \neq 0$ or there is a point p' at which $g(p') \neq 0$. Say it is the former. Then, by continuity, $f(p) \neq 0$ on an entire disc centered at p. But then it follows, since $f \cdot g \equiv 0$, that $g \equiv 0$ on that disc. Thus it must be, by the remarks in the first paragraph of this subsection, that $g \equiv 0$.

3. We have the following powerful result:

> Let $U \subseteq \mathbb{C}$ be connected and open and let f be holomorphic on U. If there is a $P \in U$ such that
>
> $$\left(\frac{\partial}{\partial z}\right)^j f(P) = 0$$
>
> for every $j \in \{0, 1, 2, \dots\}$, then $f \equiv 0$.

The reason for this result is simplicity itself: The power series expansion of f about P will have all zero coefficients. Since the series certainly converges to f on some small disc centered at P, the function is identically equal to 0 on that disc. Now, by our uniqueness result for zero sets, we conclude that f is identically 0.

4. If f and g are entire holomorphic functions and if $f(x) = g(x)$ for all $x \in \mathbb{R} \subseteq \mathbb{C}$, then $f \equiv g$. It also holds that functional identities that are true for all real values of the variable are also true for complex values of the variable (Figure 5.5). For instance,

$$\sin^2 z + \cos^2 z = 1 \qquad \text{for all} \;\; z \in \mathbb{C} \qquad (5.26)$$

because the identity is true for all $z = x \in \mathbb{R}$. This is an instance of the "principle of persistence of functional relations"—see [GRK].

Of course these statements are true because, if U is a connected open set having nontrivial intersection with the x-axis and if f holomorphic on U vanishes on that intersection, then the zero set certainly has an interior accumulation point. Again see Figure 5.5.

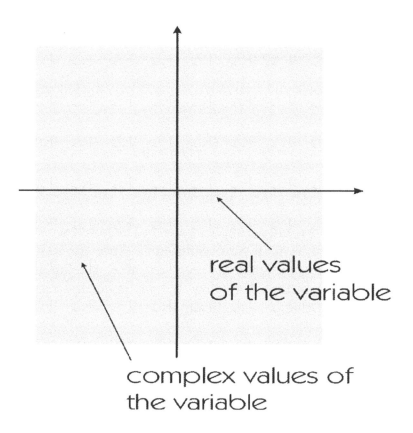

Figure 5.5: The principle of persistence of functional relations.

Exercises

1. Let f and g be entire functions and suppose that $f(x+ix^2) = g(x+ix^2)$ whenever x is real. Show that $f(z) = g(z)$ for all z.

2. Let $P_j \in D$ be defined by $P_j = 1 - 1/j$, $j = 1, 2, \dots$. Suppose that f and g are holomorphic on the disc D and that $f(P_j) = g(P_j)$ for every j. Does it follow that $f \equiv g$?

3. The real axis cannot be the zero set of a not-identically-zero holomorphic function on the entire plane. But it *can* be the zero set of a not-identically-zero harmonic function on the plane. Demonstrate both of these statements.

4. Give an example of a holomorphic function on the disc D that vanishes *on an infinite set in D* but which is not identically zero.

5. Let f be a holomorphic function on the disc D and let g be a holomorphic function on D. Let $\mathcal{P}$ be the zero set of f and let $\mathcal{Q}$ be the zero set of g. Is $\mathcal{P} \cup \mathcal{Q}$ the zero set of some holomorphic function on D? Is $\mathcal{P} \cap \mathcal{Q}$ the zero set of some holomorphic function on D? Is $\mathcal{P} \setminus \mathcal{Q}$ the zero set of some holomorphic function on D?

6. Give an example of an entire function that vanishes at every point of the form $0 + ik$ and every point of the form $k + i0$, for $k \in \mathbb{Z}$.

7. Let $c \in C$ satisfy $|c| < 1$. The function

$$\varphi_c(z) \equiv \frac{z - c}{1 - \bar{c}z}$$

is called a *Blaschke factor* at the point c. Verify these properties of φ_c:

- $|\varphi_c(z)| = 1$ whenever $|z| = 1$;
- $|\varphi_c(z)| < 1$ whenever $|z| < 1$;
- $\varphi_c(c) = 0$;
- $\varphi_c \circ \varphi_{-c}(z) \equiv z$.

8. Give an example of a holomorphic function on $\Omega \equiv D \setminus \{0\}$ such that $f(1/j) = 0$ for $j = \pm 1, \pm 2, \dots$, yet f is not identically 0.

9.   Show that a not-identically-zero entire function f cannot have uncountably many zeros.

10.   Suppose that f is a holomorphic function on the disc and $f(z)/z \equiv 1$ for z real (with the meaning of this statement for $z = 0$ suitably interpreted). What can you conclude about f?

11.   Let f, g be holomorphic on the disc D and suppose that $[f \cdot g](z) = 0$ for $z = 1/2, 1/3, 1/4, \ldots$. Show that either $f \equiv 0$ or $g \equiv 0$.

Chapter 6

Isolated Singularities and Laurent Series

6.1 The Behavior of a Holomorphic Function near an Isolated Singularity

6.1.1 Isolated Singularities

It is often important to consider a function that is holomorphic on a punctured open set $U \setminus \{P\} \subset \mathbb{C}$. Refer to Figure 6.1.

In this chapter we shall obtain a new kind of infinite series expansion which generalizes the idea of the power series expansion of a holomorphic function about a (nonsingular) point—see §§5.1.6. We shall in the process completely classify the behavior of holomorphic functions near an isolated singular point (§§6.1.3).

6.1.2 A Holomorphic Function on a Punctured Domain

Let $U \subseteq \mathbb{C}$ be an open set and $P \in U$. Suppose that $f : U \setminus \{P\} \to \mathbb{C}$ is holomorphic. In this situation we say that f has an *isolated singular point* (or *isolated singularity*) at P. The implication of the phrase is usually just that f is defined and holomorphic on some such "deleted neighborhood" of P. The specification of the set U is of secondary interest; we wish to consider the behavior of f "near P."

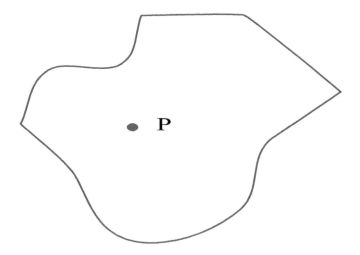

Figure 6.1: A punctured domain.

6.1.3   Classification of Singularities

There are three possibilities for the behavior of f near P that are worth distinguishing:

1. $|f(z)|$ is bounded on $D(P,r) \setminus \{P\}$ for some $r > 0$ with $D(P,r) \subseteq U$; i.e., there is some $r > 0$ and some $M > 0$ such that $|f(z)| \leq M$ for all $z \in U \cap D(P,r) \setminus \{P\}$.

2. $\lim_{z \to P} |f(z)| = +\infty$.

3.   Neither 1 nor 2.

Clearly these three possibilities cover all possible situations. It is our job now to identify extrinsically what each of these three situations entails.

6.1.4   Removable Singularities, Poles, and Essential Singularities

We shall see momentarily that, if case **1** holds, then f has a limit at P that extends f so that it is holomorphic on all of U. It is commonly said in this circumstance that f has a *removable singularity* at P. In case **2**, we will say that f has a *pole* at P. In case **3**, the function f will be said to have an *essential singularity* at P. Our goal in this and the next subsection is to understand **1**–**3** in some further detail.

6.1.5 The Riemann Removable Singularities Theorem

Let $f : D(P, r) \setminus \{P\} \to \mathbb{C}$ be holomorphic and bounded. Then

(a) $\lim_{z \to P} f(z)$ exists.

(b) The function $\widehat{f} : D(P, r) \to \mathbb{C}$ defined by

$$\widehat{f}(z) = \begin{cases} f(z) & \text{if} \quad z \neq P \\ \lim_{\zeta \to P} f(\zeta) & \text{if} \quad z = P \end{cases}$$

is holomorphic.

The reason that this theorem is true is the following. We may assume without loss of generality—by a simple translation of coordinates—that $P = 0$. Now consider the auxiliary function $g(z) = z^2 \cdot f(z)$. Then one may verify by direct application of the derivative that g is continuously differentiable at all points—including the origin. Furthermore, we may calculate with $\partial/\partial \overline{z}$ to see that g satisfies the Cauchy–Riemann equations. Thus g is holomorphic. But the very definition of g shows that g vanishes to order 2 at 0. Thus the power series expansion of g about 0 cannot have a constant term and cannot have a linear term. It follows that

$$g(z) = a_2 z^2 + a_3 z^3 + a_4 z^4 + \cdots = z^2(a_2 + a_3 z + a_4 z^2 + \cdots) \equiv z^2 \cdot h(z).$$

Notice that the function h is holomorphic—we have in fact given its power series expansion explicitly. But now, for $z \neq 0$, $h(z) = g(z)/z^2 = f(z)$. Thus we see that h is the holomorphic continuation of f (across the singularity at 0) that we seek.

6.1.6   The Casorati–Weierstrass Theorem

THEOREM 1 *If $f : D(P, r_0) \setminus \{P\} \to \mathbb{C}$ is holomorphic and P is an essential singularity of f, then $f(D(P, r) \setminus \{P\})$ is dense in $\mathbb{C}$ for any $0 < r < r_0$.*

The verification of this result is a nice application of the Riemann removable singularities theorem. For suppose to the contrary that $f(D(P, r) \setminus \{P\})$ is *not* dense in $\mathbb{C}$. This means that there is a disc $D(Q, s)$ that is *not* in the range of f. So consider the function

$$g(z) = \frac{1}{f(z) - Q}.$$

We see that the denominator of this function is bounded away from 0 (by s) hence the function g itself is bounded near P. So we may apply Riemann's theorem and conclude that g continues analytically across the point P. And the value of g near P cannot be 0. But then it follows that

$$f(z) = \frac{1}{g(z)} + Q$$

extends analytically across P. That contradicts the hypothesis that P is an essential singularity for f.

6.1.7   Concluding Remarks

Now we have seen that, at a removable singularity P, a holomorphic function f on $D(P, r_0) \setminus \{P\}$ can be continued to be holomorphic on all of $D(P, r_0)$. And, near an essential singularity at P, a holomorphic function g on $D(P, r_0) \setminus \{P\}$ has image that is dense in $\mathbb{C}$. The third possibility, that h has a *pole* at P, has yet to be described. Suffice it to say that, at a pole (case **2**), the limit of the modulus of the function is $+\infty$ hence the graph of the modulus of the function looks like a pole! See Figure 6.2. This case will be examined further in the next section.

We next develop a new type of doubly infinite series that will serve as a tool for understanding isolated singularities—especially poles.

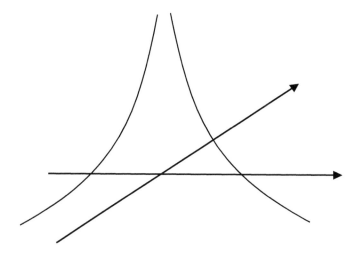

Figure 6.2: A pole.

Exercises

1. Discuss the singularities of these functions at 0:

   (a) $f(z) = \dfrac{z^2}{1 - \cos z}$

   (b) $g(z) = \dfrac{\sin z}{z}$

   (c) $h(z) = \dfrac{\sec z - 1}{\sin^2 z}$

   (d) $f(z) = \dfrac{\log(1 + z)}{z^2}$

   (e) $g(z) = \dfrac{z^2}{e^z - 1}$

   (f) $h(z) = e^{1/z}$

2. If f has a pole at P and g has a pole at P does it then follow that $f \cdot g$ has a pole at P? How about $f + g$?

3. If f has a pole at P and g has an essential singularity at P does it then follow that $f \cdot g$ has an essential singularity at P? How about $f + g$?

4.   Suppose that f is holomorphic in a deleted neighborhood $D(P,r)\setminus\{P\}$ of P and that f is not bounded near P. Assume further that $(z-P)^2 \cdot f$ *is* bounded (near P). Show that f has a pole at p. What happens if the exponent 2 is replaced by some other positive integer?

5.   Suppose that f is holomorphic in a deleted neighborhood $D(P,r)\setminus\{P\}$ of P and that $(z-P)^k \cdot f$ is unbounded for very choice of positive integer k. What conclusion can you draw about the singularity of f at P?

6.   In the Riemann removable singularities theorem, the hypothesis of boundedness is not essential. Describe a weaker hypothesis that will give (with the same reasoning!) the same conclusion. [**Hint:** Consider square integrability.]

6.2   Expansion around Singular Points

6.2.1   Laurent Series

A *Laurent series* on $D(P,r)$ is a (formal) expression of the form

$$\sum_{j=-\infty}^{+\infty} a_j(z-P)^j. \tag{6.6}$$

Note that the series sums from $j=-\infty$ to $j=\infty$ and that the individual terms are each defined for all $z \in D(P,r)\setminus\{P\}$.

6.2.2   Convergence of a Doubly Infinite Series

To discuss convergence of Laurent series, we must first make a general agreement as to the meaning of the convergence of a "doubly infinite" series $\sum_{j=-\infty}^{+\infty} \alpha_j$. We say that such a series *converges* if $\sum_{j=0}^{+\infty} \alpha_j$ and $\sum_{j=1}^{+\infty} \alpha_{-j} = \sum_{j=-\infty}^{-1} \alpha_j$ converge in the usual sense. In this case, we set

$$\sum_{-\infty}^{+\infty} \alpha_j = \left(\sum_{j=0}^{+\infty} \alpha_j\right) + \left(\sum_{j=1}^{+\infty} \alpha_{-j}\right). \tag{6.7}$$

Thus a doubly infinite series converges precisely when the sum of its "positive part" (i.e., the terms of positive index) converges and the sum of its "negative part" (i.e., the terms of negative index) converges.

We can now present the analogues for Laurent series of our basic results about power series.

6.2.3 Annulus of Convergence

The set of convergence of a Laurent series is either an open set of the form $\{z : 0 \leq r_1 < |z - P| < r_2\}$, together with perhaps some or all of the boundary points of the set, *or* a set of the form $\{z : 0 \leq r_1 < |z - P| < +\infty\}$, together with perhaps some or all of the boundary points of the set. Such an open set is called an *annulus* centered at P. We shall let

$$D(P, +\infty) = \{z : |z - P| < +\infty\} = \mathbb{C},$$

$$D(P, 0) = \{z : |z - P| < 0\} = \emptyset,$$

and

$$\overline{D}(P, 0) = \{P\}.$$

As a result, all (open) annuli (plural of "annulus") can be written in the form

$$D(P, r_2) \setminus \overline{D}(P, r_1), \quad 0 \leq r_1 \leq r_2 \leq +\infty. \tag{6.11}$$

In precise terms, the "domain of convergence" of a Laurent series is given as follows:

Let

$$\sum_{j=-\infty}^{+\infty} a_j (z - P)^j \tag{6.12}$$

be a doubly infinite series. There are (see (6.7), (6.11)) unique nonnegative extended real numbers r_1 and r_2 (r_2 may be $+\infty$) such that the series converges absolutely for all z with $r_1 < |z - P| < r_2$ and diverges for z with $|z - P| < r_1$ or $|z - P| > r_2$. Also, if $r_1 < s_1 \leq s_2 < r_2$, then $\sum_{j=-\infty}^{+\infty} |a_j(z - P)^j|$ converges uniformly on $\{z : s_1 \leq |z - P| \leq s_2\}$ and, consequently, $\sum_{j=-\infty}^{+\infty} a_j(z - P)^j$ converges absolutely and uniformly there.

6.2.4 Uniqueness of the Laurent Expansion

Let $0 \leq r_1 < r_2 \leq \infty$. If the Laurent series $\sum_{j=-\infty}^{+\infty} a_j(z-P)^j$ converges on $D(P,r_2) \setminus \overline{D}(P,r_1)$ to a function f, then, for any r satisfying $r_1 < r < r_2$, and each $j \in \mathbb{Z}$,

$$a_j = \frac{1}{2\pi i} \oint_{|\zeta-P|=r} \frac{f(\zeta)}{(\zeta-P)^{j+1}} \, d\zeta. \qquad (6.13)$$

In particular, the a_j's are uniquely determined by f.

We turn now to establishing that convergent Laurent expansions of functions holomorphic on an annulus do in fact exist.

6.2.5 The Cauchy Integral Formula for an Annulus

Suppose that $0 \leq r_1 < r_2 \leq +\infty$ and that $f : D(P,r_2) \setminus \overline{D}(P,r_1) \to \mathbb{C}$ is holomorphic. Then, for each s_1, s_2 such that $r_1 < s_1 < s_2 < r_2$ and each $z \in D(P,s_2) \setminus \overline{D}(P,s_1)$, it holds that

$$f(z) = \frac{1}{2\pi i} \oint_{|\zeta-P|=s_2} \frac{f(\zeta)}{\zeta-z} \, d\zeta - \frac{1}{2\pi i} \oint_{|\zeta-P|=s_1} \frac{f(\zeta)}{\zeta-z} \, d\zeta. \qquad (6.14)$$

The easiest way to confirm the validity of this formula is to use a little manipulation of the Cauchy formula that we already know. Examine Figure 6.3. It shows a classical Cauchy contour for a holomorphic function with *no singularity* on a neighborhood of the curve and its interior. Now we simply let the two vertical edges coalesce to form the Cauchy integral over two circles as in Figure 6.3.

6.2.6 Existence of Laurent Expansions

Now we have our main result:

If $0 \leq r_1 < r_2 \leq \infty$ and $f : D(P,r_2) \setminus \overline{D}(P,r_1) \to \mathbb{C}$ is holomorphic, then there exist complex numbers a_j such that

$$\sum_{j=-\infty}^{+\infty} a_j(z-P)^j \qquad (6.15)$$

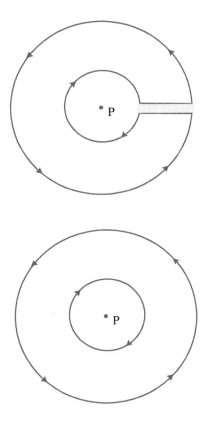

Figure 6.3: The Cauchy integral for an isolated singularity.

converges on $D(P, r_2) \setminus \overline{D}(P, r_1)$ to f. If $r_1 < s_1 < s_2 < r_2$, then the series converges absolutely and uniformly on $\overline{D}(P, s_2) \setminus D(P, s_1)$.

The series expansion is independent of s_1 and s_2. In fact, for each fixed $j = 0, \pm 1, \pm 2, \ldots$, the value of

$$a_j = \frac{1}{2\pi i} \oint_{|\zeta - P| = r} \frac{f(\zeta)}{(\zeta - P)^{j+1}} \, d\zeta \qquad (6.16)$$

is independent of r provided that $r_1 < r < r_2$.

We may justify the Laurent expansion in the following manner.

If $0 \le r_1 < s_1 < |z - P| < s_2 < r_2$, then the two integrals on the right-hand side of the equation in (6.14) can each be expanded in a series. For the first integral we have

$$\oint_{|\zeta - P| = s_2} \frac{f(\zeta)}{\zeta - z} \, d\zeta = \oint_{|\zeta - P| = s_2} \frac{f(\zeta)}{1 - \frac{z - P}{\zeta - P}} \cdot \frac{1}{\zeta - P} \, d\zeta$$

$$= \oint_{|\zeta - P| = s_2} \frac{f(\zeta)}{\zeta - P} \sum_{j=0}^{+\infty} \frac{(z - P)^j}{(\zeta - P)^j} \, d\zeta$$

$$= \oint_{|\zeta - P| = s_2} \sum_{j=0}^{+\infty} \frac{f(\zeta)(z - P)^j}{(\zeta - P)^{j+1}} \, d\zeta$$

where the geometric series expansion of

$$\frac{1}{1 - (z - P)/(\zeta - P)}$$

converges because $|z - P|/|\zeta - P| = |z - P|/s_2 < 1$. In fact, since the value of $|(z - P)/(\zeta - P)|$ is independent of ζ, for $|\zeta - P| = s_2$, it follows that the geometric series converges uniformly.

Thus we may switch the order of summation and integration to obtain

$$\oint_{|\zeta - P| = s_2} \frac{f(\zeta)}{\zeta - z} \, d\zeta = \sum_{j=0}^{+\infty} \left(\oint_{|\zeta - P| = s_2} \frac{f(\zeta)}{(\zeta - P)^{j+1}} \, d\zeta \right) (z - P)^j.$$

For $s_1 < |z - P|$, similar arguments justify the formula

$$
\oint_{|\zeta - P| = s_1} \frac{f(\zeta)}{\zeta - z} \, d\zeta = -\oint_{|\zeta - P| = s_1} \frac{f(\zeta)}{1 - \frac{\zeta - P}{z - P}} \cdot \frac{1}{z - P} \, d\zeta
$$

$$
= -\oint_{|\zeta - P| = s_1} \frac{f(\zeta)}{z - P} \sum_{j=0}^{+\infty} \frac{(\zeta - P)^j}{(z - P)^j} \, d\zeta
$$

$$
= -\sum_{j=0}^{+\infty} \left[\oint_{|\zeta - P| = s_1} f(\zeta) \cdot (\zeta - P)^j \, d\zeta \right] (z - P)^{-j-1}
$$

$$
= -\sum_{j=-\infty}^{j=-1} \left[\oint_{|\zeta - P| = s_1} \frac{f(\zeta)}{(\zeta - P)^{j+1}} \, d\zeta \right] (z - P)^j .
$$

Thus

$$
2\pi i f(z) = \sum_{j=-\infty}^{j=-1} \left[\oint_{|\zeta - P| = s_1} \frac{f(\zeta)}{(\zeta - P)^{j+1}} \, d\zeta \right] (z - P)^j
$$

$$
+ \sum_{j=0}^{+\infty} \left[\oint_{|\zeta - P| = s_2} \frac{f(\zeta)}{(\zeta - P)^{j+1}} \, d\zeta \right] (z - P)^j ,
$$

as desired.

Certainly one of the important benefits of the reasoning we have just presented is that we have an explicit formula for the coefficients of the Laurent expansion:

$$
a_j = \frac{1}{2\pi i} \oint_{\partial D(P,s)} \frac{f(\zeta)}{(\zeta - P)^{j+1}} \, d\zeta, \quad \text{any } 0 < s < r .
$$

In Subsection 7.1.2 we shall give an even more practical means, with examples, for the calculation of Laurent coefficients.

6.2.7 Holomorphic Functions with Isolated Singularities

Now let us specialize what we have learned about Laurent series expansions to the case of $f : D(P, r) \setminus \{P\} \to \mathbb{C}$ holomorphic, that is, to a holomorphic function with an isolated singularity. Thus we will be considering the Laurent expansion on a degenerate annulus of the form $D(P,r) \setminus \overline{D}(P,0)$.

If $f : D(P,r) \setminus \{P\} \to \mathbb{C}$ is holomorphic, then f has a unique Laurent series expansion

$$f(z) = \sum_{j=-\infty}^{\infty} a_j (z - P)^j \qquad (6.17)$$

that converges absolutely for $z \in D(P,r) \setminus \{P\}$. The convergence is uniform on compact subsets of $D(P,r) \setminus \{P\}$. The coefficients are given by

$$a_j = \frac{1}{2\pi i} \oint_{\partial D(P,s)} \frac{f(\zeta)}{(\zeta - P)^{j+1}} \, d\zeta, \quad \text{any } 0 < s < r. \qquad (6.18)$$

6.2.8   Classification of Singularities in Terms of Laurent Series

There are three mutually exclusive possibilities for the Laurent series

$$\sum_{j=-\infty}^{\infty} a_j (z - P)^j \qquad (6.19)$$

about an isolated singularity P:

(6.20) $a_j = 0$ for all $j < 0$.

(6.21) For some $k \geq 1, a_j = 0$ for all $-\infty < j < -k$, but $a_{-k} \neq 0$.

(6.22) Neither **(i)** nor **(ii)** applies.

These three cases correspond exactly to the three types of isolated singularities that we discussed in §§6.1.3: case **(6.20)** occurs if and only if P is a removable singularity; case **(6.21)** occurs if and only if P is a pole (of order k); and case **(6.22)** occurs if and only if P is an essential singularity.

To put this matter in other words: In case **(20)**, we have a power series that converges, of course, to a holomorphic function. In case **(21)**, our Laurent series has the form

$$\sum_{j=-k}^{\infty} a_j (z-P)^j = (z-P)^{-k} \sum_{j=-k}^{\infty} a_j (z-P)^{j+k} = (z-P)^{-k} \sum_{j=0}^{\infty} a_{j-k} (z-P)^j.$$

$$(6.20)$$

Since $a_{-k} \neq 0$, we see that, for z near P, the function defined by the series behaves like $a_{-k} \cdot (z - P)^{-k}$. In short, the function (in absolute value) blows up like $|z - P|^{-k}$ as $z \to P$. A graph in $(|z|, |f(z)|)$-space would exhibit a "pole-like" singularity. This is the source of the terminology "pole". See Figure 6.2. Case **(22)**, corresponding to an essential singularity, is much more complicated; in this case there are infinitely many negative terms in the Laurent expansion and, by Casorati–Weierstrass (§§6.1.6), they interact in a complicated fashion.

Exercises

1. Derive the Laurent expansion for the function $g(z) = e^{1/z}$ about $z = 0$. Use your knowledge of the exponential function plus substitution.

2. Derive the Laurent expansion for the function $h(z) = \frac{\sin z}{z^3}$ about $z = 0$.

3. Derive the Laurent expansion for the function $f(z) = \frac{\sin z}{\cos z}$ about $z = \pi/2$. Use long division.

4. Verify that the functions
$$f(z) = e^{1/z}$$
and
$$g(z) = \cos(1/z)$$
each have an essential singularity at $z = 0$. Now determine the nature of the behavior of f/g at 0.

5. Suppose that the function f has an essential singularity at 0. Does it then follow that $1/f$ has an essential singularity at 0?

6. Explain using Laurent series why f and g could both have essential singularities at P yet $f - g$ may not have such a singularity at P. Does a similar analysis apply to $f \cdot g$?

7. Explain using Laurent series why f and g could both have poles at P yet $f - g$ may not have such a singularity at P. Does a similar analysis apply to $f \cdot g$?

8. Given an example of functions f and g, each of which has an essential singularity at 0, yet $f + g$ has a pole of order 1 at 0.

Chapter 7

Meromorphic Functions and Laurent Expansions

7.1 Examples of Laurent Expansions

7.1.1 Principal Part of a Function

When f has a pole at P, it is customary to call the negative power part of the Laurent expansion of f around P the *principal part* of f at P. (Occasionally we shall also use the terminology "Laurent polynomial.") That is, if

$$f(z) = \sum_{j=-k}^{\infty} a_j (z-P)^j \tag{7.1}$$

for z near P, then the *principal part* of f at P is

$$\sum_{j=-k}^{-1} a_j (z-P)^j. \tag{7.2}$$

EXAMPLE 1 The Laurent expansion about 0 of the function $f(z) = (z^2 +$

$1)/\sin(z^3)$ is

$$
\begin{aligned}
f(z) &= (z^2+1) \cdot \frac{1}{\sin(z^3)} \\
&= (z^2+1) \cdot \frac{1}{z^3} \cdot \frac{1}{1 - z^6/3! + \cdots} \\
&= \frac{1}{z^3} + \frac{1}{z} + \text{ (a holomorphic function)}.
\end{aligned}
$$

The principal part of f is $1/z^3 + 1/z$. ☐

EXAMPLE 2 For a second example, consider the function $f(z) = (z^2 + 2z + 2)\sin(1/(z+1))$. Its Laurent expansion about the point -1 is

$$
\begin{aligned}
f(z) &= ((z+1)^2 + 1) \cdot \left[\frac{1}{z+1} - \frac{1}{6(z+1)^3} + \frac{1}{120(z+1)^5} \right. \\
&\qquad \left. - \frac{1}{5040(z+1)^7} + - \cdots \right] \\
&= (z+1) + \frac{5}{6}\frac{1}{(z+1)} - \frac{19}{120}\frac{1}{(z+1)^3} + \frac{41}{5040}\frac{1}{(z+1)^5} - + \cdots .
\end{aligned}
$$

The principal part of f at the point -1 is

$$
\frac{5}{6}\frac{1}{(z+1)} - \frac{19}{120}\frac{1}{(z+1)^3} + \frac{41}{5040}\frac{1}{(z+1)^5} - + \cdots . \tag{7.3}
$$

☐

7.1.2 Algorithm for Calculating the Coefficients of the Laurent Expansion

Let f be holomorphic on $D(P,r) \setminus \{P\}$ and suppose that f has a pole of order k at P. Then the Laurent series coefficients a_j of f expanded about the point P, for $j = -k, -k+1, -k+2, \ldots$, are given by the formula

$$
a_j = \frac{1}{(k+j)!} \left(\frac{\partial}{\partial z} \right)^{k+j} \left((z-P)^k \cdot f \right) \Big|_{z=P}. \tag{7.4}
$$

EXAMPLE 3 Let $f(z) = \cot z$. Let us calculate the Laurent coefficients of negative index for f at the point $P = 0$.

We first notice that
$$\cot z = \frac{\cos z}{\sin z} \, .$$
Since $\cos z = 1 - z^2/2! + - \cdots$ and $\sin z = z - z^3/3! + - \cdots$, we see immediately that, for $|z|$ small, $\cot z = \cos z / \sin z \approx 1/z$ so that f has a pole of order 1 at 0. Thus, in our formula for the Laurent coefficients, $k = 1$. Also the only Laurent coefficient of negative index is $j = -1$.

Now we see that
$$a_{-1} = \frac{1}{0!} \left(\frac{\partial}{\partial z} \right)^0 \left(z \cdot \frac{\cos z}{\sin z} \right) \bigg|_{z=0} = \left(z \cdot \frac{\cos z}{\sin z} \right) \bigg|_{z=0} \, .$$

It is appropriate to apply l'Hôpital's Rule to evaluate this last expression. Thus we have
$$\lim_{z \to 0} \frac{\cos z - z \cdot \sin z}{\cos z} = 1 \, .$$

Not surprisingly, we find that the "pole" term of the Laurent expansion of this function f about 0 is $1/z$. We say "not surprisingly" because $\cos z = 1 - + \cdots$ and $\sin z = z - + \cdots$ and hence we expect that $\cot z = 1/z + \cdots$. □

We invite the reader to use the technique of the last example to calculate a_0 for the given function f. Of course you will find l'Hôpital's Rule useful. You should not be surprised to learn that $a_0 = 0$ (and we say "not surprised" because you could have anticipated this result using long division).

EXAMPLE 4 Let us use formula (7.4) to calculate the negative Laurent coefficients of the function $g(z) = z^2/(z-1)^2$ at the point $P = 1$.

It is clear that the pole at $P = 1$ has order $k = 2$. Thus we calculate
$$a_{-2} = \frac{1}{0!} \left(\frac{\partial}{\partial z} \right)^0 \left((z-1)^2 \cdot \frac{z^2}{(z-1)^2} \right) \bigg|_{z=1} = z^2 \bigg|_{z=1} = 1$$
and
$$a_{-1} = \frac{1}{1!} \left(\frac{\partial}{\partial z} \right)^1 \left((z-1)^2 \cdot \frac{z^2}{(z-1)^2} \right) \bigg|_{z=1} = \frac{\partial}{\partial z} z^2 \bigg|_{z=1} = 2z \bigg|_{z=1} = 2 \, .$$

Of course this result may be derived by more elementary means, using just algebra:

$$\frac{z^2}{(z-1)^2} = \frac{(z-1)^2}{(z-1)^2} + \frac{2z-1}{(z-1)^2} = 1 + \frac{2z-2}{(z-1)^2} + \frac{1}{(z-1)^2} = 1 + \frac{2}{z-1} + \frac{1}{(z-1)^2}.$$

$\square$

The justification for Formula (7.4) is simplicity itself. Suppose that f has a pole of order k at the point P. We may write

$$f(z) = (z-P)^{-k} \cdot h(z),$$

where h is holomorphic near P. Writing out the ordinary power series expansion of h, we find that

$$
\begin{aligned}
f(z) &= (z-P)^{-k} \cdot \left(a_0 + a_1(z-P) + a_2(z-P)^2 + \cdots \right) \\
&= \frac{a_0}{(z-P)^k} + \frac{a_1}{(z-P)^{k-1}} + \frac{a_2}{(z-P)^{k-2}} + \cdots.
\end{aligned}
$$

So the $-(k-j)^{\text{th}}$ Laurent coefficient of f is just the same as the jth power series coefficient of h. That is the key to our calculation, because

$$h(z) = (z-P)^k \cdot f(z),$$

and thus formula (5.26) is immediate.

Exercises

1. Calculate the Laurent series of the function $f(z) = \frac{z+\sin z}{\cos z}$ at $z = \pi/2$.

2. Calculate the Laurent series of the function $g(z) = \frac{\ln z}{(z-1)^3}$ about the point $z = 1$.

3. Calculate the Laurent series of the function $\sin(1/z)$ about the point $z = 0$.

4. Calculate the Laurent series of the function $\tan z$ about the points $z = 0$, $z = \pi/2$ and $z = \pi$.

5. Suppose that f has a pole of order 1 at $z = 0$. What can you say about the behavior of $g(z) = e^{f(z)}$ at $z = 0$?

6. Suppose that f has an essential singularity at $z = 0$. What can you say about the behavior of $h(z) = e^{f(z)}$ at $z = 0$?

7. Let U be an open region in the plane. Let $\mathcal{M}$ denote the collection of functions on U that has a discrete set of poles and is holomorphic elsewhere (we allow the possibility that the function may have *no* poles). Explain why $\mathcal{M}$ is closed under addition, subtraction, multiplication, and division.

8. Consider Exercise 7 with the word "pole" replaced by "essential singularity". Does any part of the conclusion of that exercise still hold? Why or why not?

9. Let $P = 0$. Classify each of the following as having a removable singularity, a pole, or an essential singularity at P:

(a) $\dfrac{1}{z}$,

(b) $\sin \dfrac{1}{z}$,

(c) $\dfrac{1}{z^3} - \cos z$,

(d) $z \cdot e^{1/z} \cdot e^{-1/z^2}$,

(e) $\dfrac{\sin z}{z}$,

(f) $\dfrac{\cos z}{z}$,

(g) $\dfrac{\sum_{k=2}^{\infty} 2^k z^k}{z^3}$.

10. Show that

$$\sum_{j=1}^{\infty} 2^{-(2^j)} \cdot z^{-j}$$

converges for $z \neq 0$ and defines a function which has an essential singularity at $P = 0$.

11. A Laurent series converges on an annular region. Give examples to show that the set of convergence for a Laurent series can include some of the boundary, all of the boundary, or none of the boundary.

12. Calculate the annulus of convergence (including any boundary points) for each of the following Laurent series:

(a) $\sum_{j=-\infty}^{\infty} 2^{-j} z^j$,

(b) $\sum_{j=0}^{\infty} 4^{-j} z^j + \sum_{j=-\infty}^{-1} 3^j z^j$,

(c) $\sum_{j=1}^{\infty} z^j / j^2$,

(d) $\sum_{j=-\infty, j\neq 0}^{\infty} z^j / j^j$,

(e) $\sum_{j=-\infty}^{10} z^j / |j|!$ $(0! = 1)$,

(f) $\sum_{j=-20}^{\infty} j^2 z^j$.

13. Use formal algebra to calculate the first four terms of the Laurent series expansion of each of the following functions:

(a) $\tan z \equiv (\sin z / \cos z)$ about $\pi/2$,

(b) $e^z / \sin z$ about 0,

(c) $e^z / (1 - e^z)$ about 0,

(d) $\sin(1/z)$ about 0,

(e) $z(\sin z)^{-2}$ about 0,

(f) $z^2 (\sin z)^{-3}$ about 0.

For each of these functions, identify the type of singularity at the point about which the function has been expanded.

7.2 Meromorphic Functions and Singularities at Infinity

7.2.1 Meromorphic Functions

We have considered carefully those functions that are holomorphic on sets of the form $D(P, r) \setminus \{P\}$ or, more generally, of the form $U \setminus \{P\}$, where U is an open set in $\mathbb{C}$ and $P \in U$. As we have seen in our discussion of the calculus of residues, sometimes it is important to consider the possibility that a function could be "singular" at more than just one point. The appropriate precise definition requires a little preliminary consideration of what kinds of sets might be appropriate as "sets of singularities."

7.2.2 Discrete Sets and Isolated Points

A set S in $\mathbb{C}$ is *discrete* if and only if for each $z \in S$ there is a positive number ϵ (depending on z) such that

$$S \cap D(z, \epsilon) = \{z\}. \tag{7.5}$$

We also say in this circumstance that S consists of isolated points.

7.2.3 Definition of Meromorphic Function

Now fix an open set U; we next define the central concept of meromorphic function on U.

A *meromorphic function* f on U *with singular set* S is a function $f : U \setminus S \to \mathbb{C}$ such that

(7.79) S is discrete;

(7.80) f is holomorphic on $U \setminus S$ (note that $U \setminus S$ is necessarily open in $\mathbb{C}$);

(7.81) for each $P \in S$ and $\epsilon > 0$ such that $D(P, \epsilon) \subseteq U$ and $S \cap D(P, \epsilon) = \{P\}$, the function $f|_{D(P,\epsilon) \setminus \{P\}}$ has a (finite order) pole at P.

For convenience, one often suppresses explicit consideration of the set S and just says that f is a meromorphic function on U. Sometimes we say, informally, that a meromorphic function on U is a function on U that is holomorphic "except for poles." Implicit in this description is the idea that a pole is an "isolated singularity." In other words, a point P is a pole of f if and only if there is a disc $D(P, r)$ around P such that f is holomorphic on $D(P, r) \setminus \{P\}$ and has a pole at P. Back on the level of precise language, we see that our definition of a meromorphic function on U implies that, for each $P \in U$, either there is a disc $D(P, r) \subseteq U$ such that f is holomorphic on $D(P, r)$ *or* there is a disc $D(P, r) \subseteq U$ such that f is holomorphic on $D(P, r) \setminus \{P\}$ and has a pole at P.

7.2.4 Examples of Meromorphic Functions

Meromorphic functions are very natural objects to consider, primarily because they result from considering the (algebraic) reciprocals of holomorphic functions:

If U is a connected open set in $\mathbb{C}$ and if $f : U \to \mathbb{C}$ is a holomorphic function with $f \not\equiv 0$, then the function

$$F : U \setminus \{z : f(z) = 0\} \to \mathbb{C} \qquad (7.6)$$

defined by $F(z) = 1/f(z)$ is a meromorphic function on U with singular set (or pole set) equal to $\{z \in U : f(z) = 0\}$. In a sense that can be made precise, all meromorphic functions arise as *quotients* of holomorphic functions.

7.2.5   Meromorphic Functions with Infinitely Many Poles

It is quite possible for a meromorphic function on an open set U to have infinitely many poles in U. The function $1/\sin(1/z)$ is an obvious example on $U = D \setminus \{0\}$. Of course we can only allow the infinitely many poles to accumulate at the boundary (for the same reason that we can only allow the infinitely many zeros of a holomorphic function to accumulate at the boundary).

7.2.6   Singularities at Infinity

Our discussion so far of singularities of holomorphic functions can be generalized to include the limit behavior of holomorphic functions as $|z| \to +\infty$. This is a powerful method with many important consequences. Suppose for example that $f : \mathbb{C} \to \mathbb{C}$ is an entire function. We can associate to f a new function $G : \mathbb{C} \setminus \{0\} \to \mathbb{C}$ by setting $G(z) = f(1/z)$. The behavior of the function G near 0 reflects, in an obvious sense, the behavior of f as $|z| \to +\infty$. For instance

$$\lim_{|z| \to +\infty} |f(z)| = +\infty \qquad (7.7)$$

if and only if G has a pole at 0.

Suppose that $f : U \to \mathbb{C}$ is a holomorphic function on an open set $U \subseteq \mathbb{C}$ and that, for some $R > 0, U \supseteq \{z : |z| > R\}$. Define $G : \{z : 0 < |z| < 1/R\} \to \mathbb{C}$ by $G(z) = f(1/z)$. Then we say that

(7.8) f has a *removable singularity* at ∞ if G has a removable singularity at 0.

(7.9) f has a *pole at* ∞ if G has a pole at 0.

7.10) f has an *essential singularity* at ∞ if G has an essential singularity at 0.

EXAMPLE 5 The function $f(z) = z$ has a pole of order 1 at ∞ just because $G(z) = 1/z$ has a pole of order 1 at 0.

The function $f(z) = z^k$ has a pole of order k at ∞ (where $k = 1, 2, \ldots$) just because $G(z) = 1/z^k$ has a pole of order k at 0.

Any polynomial function of degree k has a pole of order k at ∞, by reasoning just as in the last paragraph.

The function $g(z) = e^z$ has an essential singularity at ∞ just because $G(z) = e^{1/z}$ has an essential singularity at the origin.

The function $h(z) = 1/z$ has a removable singularity at ∞ just because $G(z) = z$ has a removable singularity at the origin. □

7.2.7   The Laurent Expansion at Infinity

The Laurent expansion of G around 0, $G(z) = \sum_{-\infty}^{+\infty} a_n z^n$, yields immediately a series expansion for f which converges for $|z| > R$, namely,

$$f(z) \equiv G(1/z) = \sum_{-\infty}^{+\infty} a_n z^{-n} = \sum_{-\infty}^{+\infty} a_{-n} z^n . \qquad (7.8)$$

The series $\sum_{-\infty}^{+\infty} a_{-n} z^n$ is called the *Laurent expansion of f around* ∞. It follows from our definitions and from our earlier discussions that f has a removable singularity at ∞ if and only if the Laurent series of f at ∞ has no *positive* powers of z with non-zero coefficients. Also f has a pole at ∞ if and only if the series has only a finite number of positive powers of z with non-zero coefficients. Finally, f has an essential singularity at ∞ if and only if the series has infinitely many positive powers.

EXAMPLE 6 The function $f(z) = e^z$ has corresponding function $G(z) = e^{1/z}$. The Laurent expansion about the origin for G is

$$G(z) = \sum_{-\infty}^{0} \frac{z^j}{|j|!} .$$

It follows that the Laurent expansion for f about ∞ is

$$f(z) = \sum_{-\infty}^{0} \frac{z^{-j}}{|j|!} = \sum_{j=0}^{\infty} \frac{z^j}{j!} \,.$$

The function $g(z) = e^{1/z}$ has corresponding function $G(z) = e^z$. The Laurent expansion about the origin for G is

$$G(z) = \sum_{j=0}^{\infty} \frac{z^j}{j!} \,.$$

It follows that the Laurent expansion for f about ∞ is

$$f(z) = \sum_{j=0}^{\infty} \frac{z^{-j}}{j!} = \sum_{-\infty}^{0} \frac{z^j}{|j|!} \,.$$

The function $h(z) = z^3$ has corresponding function $G(z) = z^{-3}$. The Laurent expansion about the origin for G is

$$G(z) = \frac{1}{z^3} \,.$$

It follows that the Laurent expansion for h about ∞ is

$$h(z) = z^3 \,. \qquad \qquad \square$$

7.2.8   Meromorphic at Infinity

Let f be an entire function with a removable singularity at infinity. This means, in particular, that f is bounded near infinity. So f is bounded. But f is an entire function so it is constant.

Now suppose that f is entire and has a pole at infinity. Then $G(z) = f(1/z)$ has a pole (of some order k) at the origin. Hence $z^k G(z)$ has a removable singularity at the origin. We conclude then that $z^{-k} \cdot f(z)$ has a removable singularity at ∞.

Thus $z^{-k} \cdot f(z)$ is bounded near infinity. Certainly f is bounded on any compact subset of the plane. All told, then,

$$|f(z)| \le C(1 + |z|)^k \,.$$

Now examine the Cauchy estimates at the origin, on a disc $D(0, R)$, for the $(k + 1)^{\text{st}}$ derivative of f. We find that

$$\left| \frac{\partial^{k+1}}{\partial z^{k+1}} f(0) \right| \leq \frac{(k+1)!C(1+R)^k}{R^{k+1}} .$$

As $R \to +\infty$ we find that the $(k + 1)^{\text{st}}$ derivative of f at 0 is 0. In fact the same estimate can be demonstrated at any point P in the plane. We conclude that $f^{(k+1)} \equiv 0$. Thus f must be a polynomial of degree at most k.

We have treated the cases of an entire function f having a removable singularity or a pole at infinity. The only remaining possibility is an essential singularity at infinity.

Suppose that f is a meromorphic function defined on an open set $U \subseteq \mathbb{C}$ such that, for some $R > 0$, we have $U \supseteq \{z : |z| > R\}$. We say that f is *meromorphic* at ∞ if the function $G(z) \equiv f(1/z)$ is meromorphic in the usual sense on $\{z : |z| < 1/R\}$.

7.2.9   Meromorphic Functions in the Extended Plane

The definition of "meromorphic at ∞" as given is equivalent to requiring that, for some $R' > R$, f has no poles in $\{z \in \mathbb{C} : R' < |z| < \infty\}$ *and* that f has a pole at ∞.

A meromorphic function f on $\mathbb{C}$ which is also meromorphic at ∞ must be a rational function (that is, a quotient of polynomials in z). Conversely, every rational function is meromorphic on $\mathbb{C}$ and at ∞.

Remark: It is conventional to rephrase the ideas just presented by saying that the only functions that are meromorphic in the "extended plane" are rational functions. We will say more about the extended plane in §§11.3.1–11.3.3.

Exercises

1.   A holomorphic function f on a set of the form $\{z : |z| > R\}$, some $R > 0$, is said to have a zero at ∞ of order k if $f(1/z)$ has a zero of

order k at 0. Using this definition as motivation, give a definition of *pole* of order k at ∞. If g has a pole of order k at ∞, what property does $1/g$ have at ∞? What property does $1/g(1/z)$ have at 0?

2. This exercise develops a notion of residue at ∞.

First, note that if f is holomorphic on a set $D(0, r) \setminus \{0\}$ and if $0 < s < r$, then "the residue at 0" $= \frac{1}{2\pi i} \oint_{\partial D(0,s)} g(z) \, dz$ picks out one particular coefficient of the Laurent expansion of f about 0, namely it equals a_{-1}. If g is defined and holomorphic on $\{z : |z| > R\}$, then the residue at ∞ of g is defined to be the negative of the residue at 0 of $H(z) = z^{-2} \cdot g(1/z)$. (Because a positively oriented circle about ∞ is negatively oriented with respect to the origin and vice versa, we defined the *residue of g* at ∞ to be the *negative* of the residue of H at 0.) Show that the residue at ∞ of g is the coefficient of z in the Laurent expansion of g on $\{z : |z| > R\}$. Show also that the definition of residue of g at ∞ remains unchanged if the origin is replaced by some other point in the finite plane.

3. Refer to Exercise 2 for terminology. Let $R(z)$ be a rational function (quotient of polynomials). Show that the sum of all the residues (including the residue at ∞) of R is zero. Is this true for a more general class of functions than rational functions?

4. Refer to Exercise 2 for terminology. Calculate the residue of the given function at ∞.

(a) $f(z) = z^3 - 7z^2 + 8$

(b) $f(z) = z^2 e^z$

(c) $f(z) = (z + 5)^2 e^z$

(d) $f(z) = p(z)e^z$, for p a polynomial

(e) $f(z) = \dfrac{p(z)}{q(z)}$, where p and q are polynomials

(f) $f(z) = \sin z$

(g) $f(z) = \cot z$

(h) $f(z) = \dfrac{e^z}{p(z)}$, where p is a polynomial

Chapter 8

The Calculus of Residues and Applications

8.1 Residues

8.1.1 Functions with Multiple Singularities

It turns out to be useful, especially in evaluating various types of integrals, to consider functions that have more than one "singularity." We want to consider the following general question:

> Suppose that $f : U \setminus \{P_1, P_2, \ldots, P_n\} \to \mathbb{C}$ is a holomorphic function on an open set $U \subseteq \mathbb{C}$ with finitely many distinct points $P_1, P_2, \ldots, P_n$ removed. Suppose further that
>
> $$\gamma : [0, 1] \to U \setminus \{P_1, P_2, \ldots, P_n\} \tag{8.1}$$
>
> is a piecewise C^1 closed curve (§§4.1.3) that (typically) "surrounds" some of the points $P_1, \ldots, P_n$. Then how is $\oint_\gamma f$ related to the behavior of f near the points $P_1, P_2, \ldots, P_n$?

The first step is to restrict our attention to open sets U for which $\oint_\gamma f$ is necessarily 0 if $P_1, P_2, \ldots, P_n$ are removable singularities of f. See the next subsection.

8.1.2   The Concept of Residue

Suppose that U is a domain, $P \in U$, and f is a function holomorphic on $U \setminus \{P\}$ with a pole at P. Let γ be a simple, closed curve in U that surrounds P. And let $D(P,r)$ be a small disc, centered at P, that lies inside γ. Then certainly, by the usual Cauchy theory,

$$\frac{1}{2\pi i} \oint_{\gamma} f(z)\, dz = \frac{1}{2\pi i} \oint_{\partial D(P,r)} f(z)\, dz \,.$$

But more is true. Let a_{-1} be the -1 coefficient of the Laurent expansion of f about P. Then in fact

$$\frac{1}{2\pi i} \oint_{\gamma} f(z)\, dz = \frac{1}{2\pi i} \oint_{\partial D(P,r)} f(z)\, dz$$

$$= \frac{1}{2\pi i} \oint_{\partial D(P,r)} \frac{a_{-1}}{z-P}\, dz = a_{-1} \,. \qquad (8.2)$$

We call the value a_{-1} the *residue* of f at the point P.

The justification for formula (8.2) is the following. Observe that, with the parameterization $\mu(t) = P + re^{it}$ for $\partial D(P,r)$, we see for $j \neq -1$ that

$$\oint_{\partial D(P,r)} (z-P)^j\, dz = \int_0^{2\pi} (re^{it})^j \cdot rie^{it}\, dt = r^{j+1} i \int_0^{2\pi} e^{i(j+1)t}\, dt = 0 \,.$$

It is important in this last calculation that $j \neq -1$. If instead $j = -1$ then the integral turns out to be

$$i \int_0^{2\pi} 1\, dt = 2\pi i \,.$$

This information is critical because if we are integrating a meromorphic function $f(z) = \sum_{j=-\infty}^{\infty} a_j (z-P)^j$ around the contour $\partial D(P,r)$ then the result is

$$\oint_{\partial D(P,r)} f(z)\, dz = \oint_{\partial D(P,r)} \sum_{j=-\infty}^{\infty} a_j (z-P)^j = \sum_{j=-\infty}^{\infty} a_j \oint_{\partial D(P,r)} (z-P)^j\, dz$$

$$= a_{-1} \oint_{\partial D(P,r)} (z-P)^{-1}\, dz = 2\pi i a_{-1} \,.$$

In other words,

$$a_{-1} = \frac{1}{2\pi i} \oint_{\partial D(P,r)} f(z)\,dz.$$

We will make incisive use of this information in the succeeding subsections.

8.1.3 The Residue Theorem

Suppose that $U \subseteq \mathbb{C}$ is a simply connected open set in $\mathbb{C}$, and that $P_1, \ldots, P_n$ are distinct points of U. Suppose that $f : U \backslash \{P_1, \ldots, P_n\} \to \mathbb{C}$ is a holomorphic function and γ is a piecewise C^1 curve in $U \backslash \{P_1, \ldots, P_n\}$ that surrounds the points $P_1, P_2, \ldots, P_n$. Set

$$\begin{aligned} R_j \;\; &= \;\; \text{the coefficient of } (z - P_j)^{-1} \\ &\text{in the Laurent expansion of } f \text{ about } P_j. \end{aligned} \tag{8.3}$$

Then

$$\oint_\gamma f = \sum_{j=1}^n R_j \cdot \left(\oint_\gamma \frac{1}{\zeta - P_j} d\zeta \right). \tag{8.4}$$

The rationale behind the residue formula is straightforward from the picture. Examine Figure 8.1. It shows the curve γ and the poles $P_1, \ldots, P_n$. Figure 8.2 exhibits a small circular contour around each pole. And Figure 8.3 shows our usual trick of connecting up the contours. The integral around the big, conglomerate contour in Figure 8.3 (including γ, the integrals around each of the circular arcs, and the integrals along the connecting segments) is equal to 0. This demonstrates that

> The integral of f around γ is equal to the sum of the integrals around each of the P_j.

And that is formula (8.4).

8.1.4 Residues

The result just stated is used so often that some special terminology is commonly used to simplify its statement. First, the number R_j is usually called the *residue* of f at P_j, written $\text{Res}_f(P_j)$. Note that this

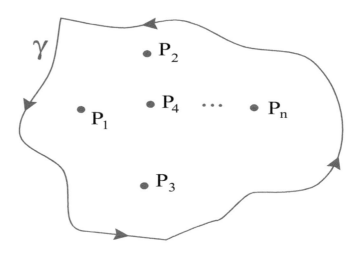

Figure 8.1: A curve γ with poles inside.

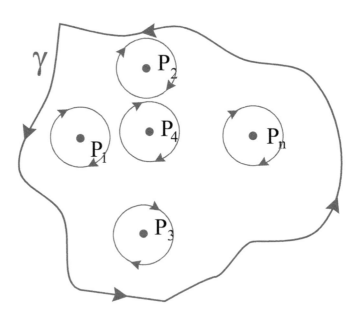

Figure 8.2: A small circle about each pole.

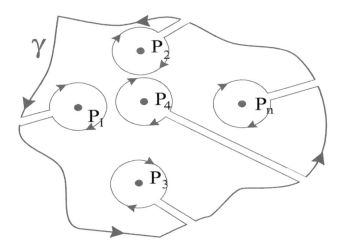

Figure 8.3: Stitching together the circles.

terminology of considering the number R_j attached to the point P_j makes sense because $\mathrm{Res}_f(P_j)$ is completely determined by knowing f in a small neighborhood of P_j. In particular, the value of the residue does not depend on what the other points P_k, $k \neq j$, might be, or on how f behaves near those points.

8.1.5 The Index or Winding Number of a Curve about a Point

The second piece of terminology associated to our result deals with the integrals that appear on the right-hand side of equation (8.4).

If $\gamma : [a, b] \to \mathbb{C}$ is a piecewise C^1 closed curve and if $P \notin \tilde{\gamma} \equiv \gamma([a, b])$, then the *index of γ with respect to P*, written $\mathrm{Ind}_\gamma(P)$, is defined to be the number

$$\frac{1}{2\pi i} \oint_\gamma \frac{1}{\zeta - P} \, d\zeta. \tag{8.5}$$

The index is also sometimes called the "winding number of the curve γ about the point P." It is a fact that $\mathrm{Ind}_\gamma(P)$ is always an integer. Figure 8.4 illustrates the index of various curves γ with respect to different points P. Intuitively, the index measures the number of times the curve wraps around P, with counterclockwise being the positive direction of wrapping and clockwise being the negative.

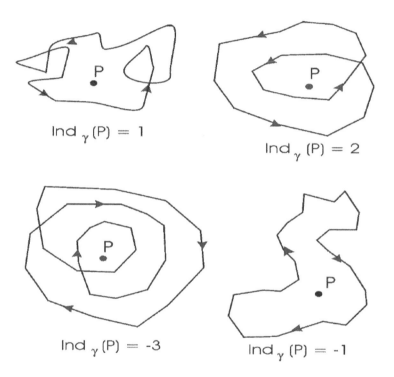

Figure 8.4: Examples of the index of a curve.

The fact that the index is an integer-valued function suggests that the index counts the topological winding of the curve γ. Note in particular that a curve that traces a circle about the origin k times in a counter-clockwise direction has index k with respect to the origin; a curve that traces a circle about the origin k times in a clockwise direction has index $-k$ with respect to the origin.

8.1.6 Restatement of the Residue Theorem

Using the notation of residue and index, the Residue Theorem's formula becomes

$$\oint_\gamma f = 2\pi i \cdot \sum_{j=1}^{n} \operatorname{Res}_f(P_j) \cdot \operatorname{Ind}_\gamma(P_j). \qquad (8.6)$$

People sometimes state this formula informally as "the integral of f around γ equals $2\pi i$ times the sum of the residues counted according to the index of γ about the singularities."

In practice, when we apply the residue theorem, we use a simple, closed curve γ. Thus the index of γ about any point in its interior is just 1. And therefore we use the ideas of Subsection 8.1.3 and replace γ with a small circle about each pole of the function (which of course will also have index equal to 1 with respect to the point at its center).

8.1.7 Method for Calculating Residues

We need a method for calculating residues.

Let f be a function with a pole of order k at P. Then

$$\operatorname{Res}_f(P) = \frac{1}{(k-1)!} \left(\frac{\partial}{\partial z}\right)^{k-1} \left((z-P)^k f(z)\right)\Big|_{z=P}. \qquad (8.7)$$

This is just a special case of the formula (7.4).

8.1.8 Summary Charts of Laurent Series and Residues

We provide two charts, the first of which summarizes key ideas about Laurent coefficients and the second of which contains key ideas about residues.

Poles and Laurent Coefficients

Item	Formula
j^{th} Laurent coefficient of f with pole of order k at P	$\dfrac{1}{(k+j)!}\dfrac{d^{k+j}}{dz^{k+j}}\left[(z-P)^k \cdot f\right]\Big|_{z=P}$
residue of f with a pole of order k at P	$\dfrac{1}{(k-1)!}\dfrac{d^{k-1}}{dz^{k-1}}\left[(z-P)^k \cdot f\right]\Big|_{z=P}$
order of pole of f at P	least integer $k \geq 0$ such that $(z-P)^k \cdot f$ is bounded near P
order of pole of f at P	$\displaystyle\lim_{z\to P}\left|\dfrac{\log|f(z)|}{\log|z-P|}\right|$

Techniques for Finding the Residue at P

Function	Type of Pole	Calculation
$f(z)$	simple	$\lim_{z \to P}(z - P) \cdot f(z)$
$f(z)$	pole of order k k is the least integer such that $\lim_{z \to P} \mu(z)$ exists, where $\mu(z) = (z - P)^k f(z)$	$\lim_{z \to P} \dfrac{\mu^{(k-1)}(z)}{(k - 1)!}$
$\dfrac{m(z)}{n(z)}$	$m(P) \neq 0,\ n(z) = 0,\ n'(P) \neq 0$	$\dfrac{m(P)}{n'(P)}$
$\dfrac{m(z)}{n(z)}$	m has zero of order k at P n has zero of order $(k + 1)$ at P	$(k + 1) \cdot \dfrac{m^{(k)}(P)}{n^{(k+1)}(P)}$
$\dfrac{m(z)}{n(z)}$	m has zero of order r at P n has zero of order $(k + r)$ at P	$\lim_{z \to P} \dfrac{\mu^{(k-1)}(z)}{(k - 1)!},$ $\mu(z) = (z - P)^k \dfrac{m(z)}{n(z)}$

Exercises

1. Calculate the residue of the function $f(z) = \cot z$ at $z = 0$.

2. Calculate the residue of the function $h(z) = \tan z$ at $z = \pi/2$.

3. Calculate the residue of the function $g(z) = e^{1/z}$ at $z = 0$.

4.   Calculate the residue of the function $f(z) = \cot^2 z$ at $z = 0$.

5.   Calculate the residue of the function $g(z) = \sin(1/z)$ at $z = 0$.

6.   Calculate the residue of the function $h(z) = \tan(1/z)$ at $z = 0$.

7.   If the function f has residue a at $z = 0$ and the function g has residue b at $z = 0$ then what can you say about the residue of f/g at $z = 0$? What about the residue of $f \cdot g$ at $z = 0$?

8.   Let f and g be as in Exercise 7. Describe the residues of $f + g$ and $f - g$ at $z = 0$.

9.   Calculate the residue at the origin of $f_k(z) = z^k$ for $k \in \mathbb{Z}$. Explain the different answers for different ranges of k.

10.   Is the residue of a function f at an essential singularity always equal to 0? Why or why not?

11.   Use the calculus of residues to compute each of the following integrals:

(a) $\dfrac{1}{2\pi i} \oint_{\partial D(0,5)} f(z)\, dz$ where $f(z) = z/[(z+1)(z+2i)]$,

(b) $\dfrac{1}{2\pi i} \oint_{\partial D(0,5)} f(z)\, dz$ where $f(z) = e^z/[(z+1)\sin z]$,

(c) $\dfrac{1}{2\pi i} \oint_{\partial D(0,8)} f(z)\, dz$ where $f(z) = \cot z/[(z-6i)^2 + 64]$,

(d) $\dfrac{1}{2\pi i} \oint_{\gamma} f(z)\, dz$ where $f(z) = \dfrac{e^z}{z(z+1)(z+2)}$ and γ is the negatively (clockwise) oriented triangle with vertices $1 \pm i$ and -3,

(e) $\dfrac{1}{2\pi i} \oint_{\gamma} f(z)\, dz$ where $f(z) = \dfrac{e^z}{(z+3i)^2(z+3)^2(z+4)}$ and γ is the negatively oriented rectangle with vertices $2 \pm i, -8 \pm i$,

(f) $\dfrac{1}{2\pi i} \oint_{\gamma} f(z)\, dz$ where $f(z) = \dfrac{\cos z}{z^2(z+1)^2(z+i)}$ and γ is as in Figure 8.5,

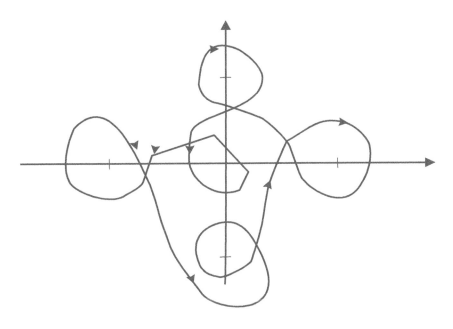

Figure 8.5: The contour in Exercise 11f.

(g) $\dfrac{1}{2\pi i} \displaystyle\oint_\gamma f(z)\,dz$ where $f(z) = \dfrac{\sin z}{z(z+2i)^3}$ and γ is as in Figure 8.6.

(h) $\dfrac{1}{2\pi i} \displaystyle\oint_\gamma f(z)\,dz$ where $f(z) = \dfrac{e^{iz}}{(\sin z)(\cos z)}$ and γ is the positively (counterclockwise) oriented quadrilateral with vertices $\pm 5i, \pm 10$.

(i) $\dfrac{1}{2\pi i} \displaystyle\oint_\gamma f(z)\,dz$ where $f(z) = \tan z$ and γ is the curve in Figure 8.7.

12. Let $R(z)$ be a rational function: $R(z) = p(z)/q(z)$ where p and q are holomorphic polynomials. Let f be holomorphic on $\mathbb{C}\setminus\{P_1, P_2, \ldots, P_k\}$ and suppose that f has a pole at each of the points $P_1, P_2, \ldots, P_k$. Finally assume that

$$|f(z)| \le |R(z)|$$

for all z at which $f(z)$ and $R(z)$ are defined. Show that f is a constant multiple of R. In particular, f is rational. [**Hint:** Think about $f(z)/R(z)$.]

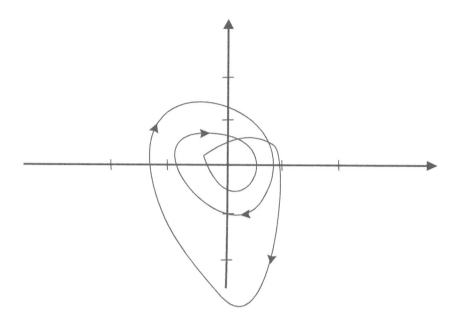

Figure 8.6: The contour in Exercise 11g.

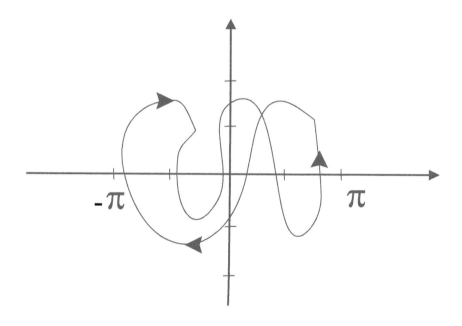

Figure 8.7: The contour in Exercise 11i.

13. Let $f : D(P, r) \setminus \{P\} \to \mathbb{C}$ be holomorphic. Let $U = f(D(P, r) \setminus \{P\})$. Assume that U is open (we shall later see that this is always the case if f is nonconstant). Let $g : U \to \mathbb{C}$ be holomorphic. If f has a removable singularity at P, does $g \circ f$ have one also? What about the case of poles and essential singularities?

8.2 Applications to the Calculation of Definite Integrals and Sums

8.2.1 The Evaluation of Definite Integrals

One of the most classical and fascinating applications of the calculus of residues is the calculation of definite (usually improper) real integrals. It is an over-simplification to call these calculations, taken together, a "technique": it is more like a *collection* of techniques. We present several instances of the method.

8.2.2 A Basic Example

To evaluate

$$\int_{-\infty}^{\infty} \frac{1}{1 + x^4} dx, \tag{8.8}$$

we "complexify" the integrand to $f(z) = 1/(1 + z^4)$ and consider the integral

$$\oint_{\gamma_R} \frac{1}{1 + z^4} dx. \tag{8.9}$$

See Figure 8.8.

Now part of the game here is to choose the right piecewise C^1 curve or "contour" γ_R. The appropriateness of our choice is justified (after the fact) by the calculation that we are about to do. Assume that $R > 1$. Define

$$\gamma_R^1(t) = t + i0 \quad \text{if} \quad -R \le t \le R,$$
$$\gamma_R^2(t) = Re^{it} \quad \text{if} \quad 0 \le t \le \pi.$$

Call these two curves, taken together, γ or γ_R.

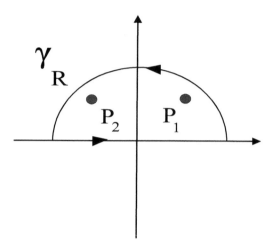

Figure 8.8: The curve γ_R in Subsection 8.2.2.

Now we set $U = \mathbb{C}$, $P_1 = 1/\sqrt{2} + i/\sqrt{2}$, $P_2 = -1/\sqrt{2} + i/\sqrt{2}$, $P_3 = -1/\sqrt{2} - i/\sqrt{2}$, $P_4 = 1/\sqrt{2} - i/\sqrt{2}$; the points P_1, P_2, P_3, P_4 are the poles of $1/[1 + z^4]$. Thus $f(z) = 1/(1 + z^4)$ is holomorphic on $U \setminus \{P_1, \ldots, P_4\}$ and the Residue Theorem applies.

On the one hand,

$$\oint_\gamma \frac{1}{1 + z^4}\, dz = 2\pi i \sum_{j=1,2} \mathrm{Ind}_\gamma(P_j) \cdot \mathrm{Res}_f(P_j),$$

where we sum only over the poles of f that lie inside γ. These are P_1 and P_2. An easy calculation shows that

$$\mathrm{Res}_f(P_1) = \frac{1}{4(1/\sqrt{2} + i/\sqrt{2})^3} = -\frac{1}{4}\left(\frac{1}{\sqrt{2}} + i\frac{1}{\sqrt{2}}\right)$$

and

$$\mathrm{Res}_f(P_2) = \frac{1}{4(-1/\sqrt{2} + i/\sqrt{2})^3} = -\frac{1}{4}\left(-\frac{1}{\sqrt{2}} + i\frac{1}{\sqrt{2}}\right).$$

Of course the index at each point is 1. So

$$\oint_\gamma \frac{1}{1 + z^4}\, dz = 2\pi i \left(-\frac{1}{4}\right)\left[\left(\frac{1}{\sqrt{2}} + i\frac{1}{\sqrt{2}}\right) + \left(-\frac{1}{\sqrt{2}} + i\frac{1}{\sqrt{2}}\right)\right]$$

$$= \frac{\pi}{\sqrt{2}}. \tag{8.10}$$

On the other hand,

$$\oint_{\gamma} \frac{1}{1+z^4}\, dz = \oint_{\gamma_R^1} \frac{1}{1+z^4}\, dz + \oint_{\gamma_R^2} \frac{1}{1+z^4}\, dz. \qquad (8.11)$$

Trivially,

$$\oint_{\gamma_R^1} \frac{1}{1+z^4}\, dz = \int_{-R}^{R} \frac{1}{1+t^4} \cdot 1 \cdot dt \to \int_{-\infty}^{\infty} \frac{1}{1+t^4}\, dt \qquad (8.12)$$

as $R \to +\infty$. That is good, because this last is the integral that we wish to evaluate. Better still,

$$\left| \oint_{\gamma_R^2} \frac{1}{1+z^4}\, dx \right| \le \{\text{length}(\gamma_R^2)\} \cdot \max_{\gamma_R^2} \left| \frac{1}{1+z^4} \right| \le \pi R \cdot \frac{1}{R^4 - 1}. \qquad (8.13)$$

[Here we use the inequality $|1 + z^4| \ge |z|^4 - 1$.] Thus

$$\left| \oint_{\gamma_R^2} \frac{1}{1+z^4}\, dz \right| \to 0 \qquad \text{as} \qquad R \to \infty. \qquad (8.14)$$

Finally, (8.10)–(8.14) taken together yield

$$\begin{aligned}
\frac{\pi}{\sqrt{2}} &= \lim_{R \to \infty} \oint_{\gamma} \frac{1}{1+z^4}\, dz \\
&= \lim_{R \to \infty} \oint_{\gamma_R^1} \frac{1}{1+z^4}\, dz + \lim_{R \to \infty} \oint_{\gamma_R^2} \frac{1}{1+z^4}\, dz \\
&= \int_{-\infty}^{\infty} \frac{1}{1+t^4}\, dt + 0.
\end{aligned}$$

This solves the problem: the value of the integral is $\pi/\sqrt{2}$.

   In other problems, it will not be so easy to pick the contour so that the superfluous parts (in the above example, this would be the integral over γ_R^2) tend to zero, nor is it always so easy to demonstrate that they *do* tend to zero. Sometimes, it is not even obvious how to complexify the integrand.

8.2.3   Complexification of the Integrand

We evaluate

$$\int_{-\infty}^{\infty} \frac{\cos x}{1 + x^2} dx \tag{8.15}$$

by using the contour γ_R as in Figure 8.8 (i.e., the same contour as in the last example). Of course the pole(s) will be different this time.

The obvious choice for the complexification of the integrand is

$$f(z) = \frac{\cos z}{1 + z^2} = \frac{[e^{iz} + e^{-iz}]/2}{1 + z^2} = \frac{[e^{ix}e^{-y} + e^{-ix}e^{y}]/2}{1 + z^2}. \tag{8.16}$$

Now $|e^{ix}e^{-y}| = |e^{-y}| \leq 1$ on γ_R but $|e^{-ix}e^{y}| = |e^{y}|$ becomes quite large on γ_R when R is large and positive. There is no evident way to alter the contour so that good estimates result. Instead, we alter the function! Let $g(z) = e^{iz}/(1 + z^2)$.

Note that, for $R > 1$, the only pole of g inside γ_R is at i. On the one hand (for $R > 1$),

$$
\begin{aligned}
\oint_{\gamma_R} g(z) &= 2\pi i \cdot \text{Res}_g(i) \cdot \text{Ind}_{\gamma_R}(i) \\
&= 2\pi i \left(\frac{1}{e(1 + i)} \right) \cdot 1 \\
&= 2\pi i \left(\frac{1 - i}{2e} \right) \\
&= \frac{\pi}{e} + \frac{\pi i}{e}.
\end{aligned}
$$

On the other hand, with $\gamma_R^1(t) = t, -R \leq t \leq R$, and $\gamma_R^2(t) = Re^{it}, 0 \leq t \leq \pi$, we have

$$\oint_{\gamma_R} g(z)\, dz = \oint_{\gamma_R^1} g(z)\, dz + \oint_{\gamma_R^2} g(z)\, dz.$$

Of course

$$\oint_{\gamma_R^1} g(z)\, dz \to \int_{-\infty}^{\infty} \frac{e^{ix}}{1 + x^2} dx \quad \text{as} \quad R \to \infty.$$

And

$$\left| \oint_{\gamma_R^2} g(z)\, dz \right| \leq \text{length}(\gamma_R^2) \cdot \max_{\gamma_R^2} |g| \leq \pi R \cdot \frac{1}{R^2 - 1} \to 0 \quad \text{as} \quad R \to \infty.$$

Thus

$$\int_{-\infty}^{\infty} \frac{\cos x}{1+x^2}\,dx = \text{Re} \int_{-\infty}^{\infty} \frac{e^{ix}}{1+x^2}\,dx = \text{Re}\left(\frac{\pi}{e} + \frac{\pi i}{e}\right) = \frac{\pi}{e}.$$

8.2.4 An Example with a More Subtle Choice of Contour

Let us evaluate

$$\int_{-\infty}^{\infty} \frac{\sin x}{x}\,dx. \tag{8.17}$$

Before we begin, we remark that $\sin x/x$ is bounded near zero; also, the integral converges at ∞ (as an improper Riemann integral) by integration by parts. So the problem makes sense. Using the lesson learned from the last example, we consider the function $g(z) = e^{iz}/z$. However, the pole of e^{iz}/z is at $z = 0$ and that lies *on the contour* in Figure 8.8. Thus *that* contour may not be used. We instead use the contour $\mu = \mu_R$ that is depicted in Figure 8.9.

Define

$$\begin{aligned}
\mu_R^1(t) &= t, & -R \leq t \leq -1/R, \\
\mu_R^2(t) &= e^{it}/R, & \pi \leq t \leq 2\pi, \\
\mu_R^3(t) &= t, & 1/R \leq t \leq R, \\
\mu_R^4(t) &= Re^{it}, & 0 \leq t \leq \pi.
\end{aligned}$$

Clearly

$$\oint_{\mu} g(z)\,dz = \sum_{j=1}^{4} \oint_{\mu_R^j} g(z)\,dz.$$

On the one hand, for $R > 0$,

$$\oint_{\mu} g(z)\,dz = 2\pi i \text{Res}_g(0) \cdot \text{Ind}_{\mu}(0) = 2\pi i \cdot 1 \cdot 1 = 2\pi i. \tag{8.18}$$

On the other hand,

$$\oint_{\mu_R^1} g(z)\,dz + \oint_{\mu_R^3} g(z)\,dz \to \int_{-\infty}^{\infty} \frac{e^{ix}}{x}\,dx \quad \text{as} \quad R \to \infty. \tag{8.19}$$

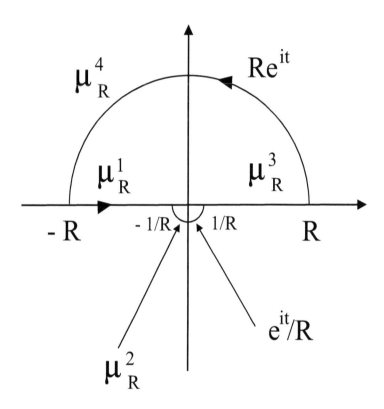

Figure 8.9: The curve μ_R in Subsection 8.2.4.

Furthermore,

$$\left| \oint_{\mu_R^4} g(z)\, dz \right| \leq \left| \oint_{\substack{\mu_R^4 \\ \operatorname{Im} z < \sqrt{R}}} g(z)\, dz \right| + \left| \oint_{\substack{\mu_R^4 \\ \operatorname{Im} z \geq \sqrt{R}}} g(z)\, dz \right|$$

$$\equiv A + B.$$

Now

$$A \leq \text{length}(\mu_R^4 \cap \{z : \operatorname{Im} z < \sqrt{R}\}) \cdot \max\{|g(z)| : z \in \mu_R^4, \operatorname{Im} z < \sqrt{R}\}$$

$$\leq 4\sqrt{R} \cdot \left(\frac{1}{R} \right) \to 0 \quad \text{as} \quad R \to \infty.$$

Also

$$B \leq \text{length}(\mu_R^4 \cap \{z : \operatorname{Im} z \geq \sqrt{R}\}) \cdot \max\{|g(z)| : z \in \mu_R^4, \operatorname{Im} z \geq \sqrt{R}\}$$

$$\leq \pi R \cdot \left(\frac{e^{-\sqrt{R}}}{R} \right) \to 0 \quad \text{as} \quad R \to \infty.$$

So

$$\left| \oint_{\mu_R^4} g(z)\, dz \right| \to 0 \quad \text{as} \quad R \to \infty. \tag{8.20}$$

Finally,

$$\oint_{\mu_R^2} g(z)\, dz = \int_\pi^{2\pi} \frac{e^{i(e^{it}/R)}}{e^{it}/R} \cdot \left(\frac{i}{R} e^{it} \right) dt$$

$$= i \int_\pi^{2\pi} e^{i(e^{it}/R)}\, dt.$$

As $R \to \infty$ this tends to

$$= i \int_\pi^{2\pi} 1\, dt$$

$$= \pi i \quad \text{as} \quad R \to \infty.$$

In summary, we see that

$$2\pi i = \oint_\mu g(z)\, dz = \sum_{j=1}^4 \oint_{\mu_R^j} g(z)\, dz$$

$$\rightarrow \int_{-\infty}^{\infty} \frac{e^{ix}}{x} dx + \pi i \quad \text{as} \quad R \to \infty.$$

Taking imaginary parts yields

$$\pi = \int_{-\infty}^{\infty} \frac{\sin x}{x} dx.$$

8.2.5 Making the Spurious Part of the Integral Disappear

Consider the integral

$$\int_0^{\infty} \frac{x^{1/3}}{1 + x^2} dx. \tag{8.21}$$

We complexify the integrand by setting $f(z) = z^{1/3}/(1 + z^2)$. Note that, on the simply connected set $U = \mathbb{C} \setminus \{iy : y \leq 0\}$, the expression $z^{1/3}$ is unambiguously defined as a holomorphic function by setting $z^{1/3} = r^{1/3}e^{i\theta/3}$ when $z = re^{i\theta}, -\pi/2 < \theta < 3\pi/2$. We again use the contour displayed in Figure 8.10.

We must do this since $z^{1/3}$ is not a well-defined holomorphic function in any neighborhood of 0. Let us use the notation from the figure. We refer to the preceding examples for some of the parameterizations that we now use.

Clearly

$$\oint_{\mu_R^3} f(z)\, dz \rightarrow \int_0^{\infty} \frac{t^{1/3}}{1 + t^2} dt. \tag{8.22}$$

Of course that is good, but what will become of the integral over μ_R^1? We have

$$\oint_{\mu_R^1} = \int_{-R}^{-1/R} \frac{t^{1/3}}{1 + t^2} dt$$

$$= \int_{1/R}^{R} \frac{(-t)^{1/3}}{1 + t^2} dt$$

$$= \int_{1/R}^{R} \frac{e^{i\pi/3}t^{1/3}}{1 + t^2} dt.$$

(by our definition of $z^{1/3}$!). Thus

$$\oint_{\mu_R^3} f(z)\,dz + \oint_{\mu_R^1} f(z)\,dz \rightarrow \left(1 + \left(\frac{1}{2} + \frac{\sqrt{3}}{2}i\right)\right)\int_0^\infty \frac{t^{1/3}}{1+t^2}dt \quad \text{as} \quad R \rightarrow \infty.$$
(8.23)

On the other hand,

$$\left|\oint_{\mu_R^4} f(z)\,dz\right| \leq \pi R \cdot \frac{R^{1/3}}{R^2 - 1} \rightarrow 0 \quad \text{as} \quad R \rightarrow \infty \qquad (8.24)$$

and

$$\begin{aligned}
\oint_{\mu_R^2} f(z)\,dz &= \int_{-\pi}^{-2\pi} \frac{(e^{it}/R)^{1/3}}{1 + e^{2it}/R^2}(i)e^{it}/R\,dt \\
&= R^{-4/3}\int_{-\pi}^{-2\pi} \frac{e^{i4t/3}}{1 + e^{2it}/R^2}\,dt \rightarrow 0 \quad \text{as} \quad R \rightarrow \infty.
\end{aligned}$$

So

$$\oint_{\mu_R} f(z)\,dz \rightarrow \left(\frac{3}{2} + \frac{\sqrt{3}}{2}i\right)\int_0^\infty \frac{t^{1/3}}{1+t^2}\,dt \quad \text{as} \quad R \rightarrow \infty. \qquad (8.25)$$

The calculus of residues tells us that, for $R > 1$,

$$\begin{aligned}
\oint_{\mu_R} f(z)\,dz &= 2\pi i \mathrm{Res}_f(i) \cdot \mathrm{Ind}_{\mu_R}(i) \\
&= 2\pi i\left(\frac{e^{i\pi/6}}{2i}\right) \cdot 1 \\
&= \pi\left(\frac{\sqrt{3}}{2} + \frac{i}{2}\right). \qquad (8.26)
\end{aligned}$$

Finally, (8.25) and (8.26) taken together yield

$$\int_0^\infty \frac{t^{1/3}}{1+t^2}\,dt = \frac{\pi}{\sqrt{3}}.$$

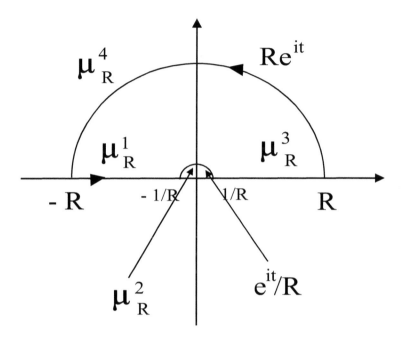

Figure 8.10: The curve η_R in Subsection 5.5.6.

8.2.6   The Use of the Logarithm

While the integral

$$\int_0^\infty \frac{dx}{x^2 + 6x + 8} \tag{8.27}$$

can be calculated using methods of calculus, it is enlightening to perform the integration by complex variable methods. Note that if we endeavor to use the integrand $f(z) = 1/(z^2 + 6z + 8)$ together with the idea of the last example, then there is no "auxiliary radius" that helps. More precisely, $((re^{i\theta})^2 + 6re^{i\theta} + 8)$ is a constant multiple of $r^2 + 6r + 8$ only if θ is an integer multiple of 2π. The following non-obvious device is often of great utility in problems of this kind. Define $\log z$ on $U \equiv \mathbb{C} \setminus \{x : x \geq 0\}$ by $\log(re^{i\theta}) = (\log r) + i\theta$ when $0 < \theta < 2\pi, r > 0$. Here $\log r$ is understood to be the standard real logarithm. Then, on U, log is a well-defined holomorphic function. [Observe here that there are infinitely many ways to define the logarithm function on U. One could set $\log(re^{i\theta}) = (\log r) + i(\theta + 2k\pi)$ for any integer choice of k. What we have done here is called "choosing a branch" of the logarithm.]

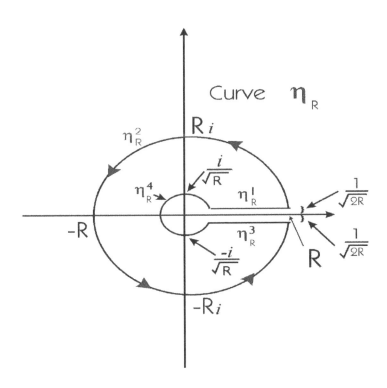

Figure 8.11: The curve Γ_n in Subsection 8.2.6.

We use the contour η_R displayed in Figure 8.11 and integrate the function $g(z) = \log z/(z^2 + 6z + 8)$. Let

$$
\begin{aligned}
\eta_R^1(t) &= t + i/\sqrt{2R}, \quad 1/\sqrt{2R} \le t \le R, \\
\eta_R^2(t) &= Re^{it}, \quad \pi/4 \le t \le 7\pi/4
\end{aligned}
$$

and

$$
\begin{aligned}
\eta_R^3(t) &= R - t - i/\sqrt{2R}, \quad 0 \le t \le R - 1/\sqrt{2R}, \\
\eta_R^4(t) &= e^{-it}/\sqrt{R}, \quad \pi/4 \le t \le 7\pi/4 \,.
\end{aligned}
$$

Now

$$
\begin{aligned}
\oint_{\eta_R} g(z)\, dz &= 2\pi i (\mathrm{Res}_{\eta_R}(-2) \cdot 1 + \mathrm{Res}_{\eta_R}(-4) \cdot 1) \\
&= 2\pi i \left(\frac{\log(-2)}{2} + \frac{\log(-4)}{-2} \right) \\
&= 2\pi i \left(\frac{\log 2 + \pi i}{2} + \frac{\log 4 + \pi i}{-2} \right) \\
&= -\pi i \log 2 \,.
\end{aligned}
\tag{8.28}
$$

Also, it is straightforward to check that

$$
\left| \oint_{\eta_R^2} g(z)\, dz \right| \to 0,
\tag{8.29}
$$

$$
\left| \oint_{\eta_R^4} g(z)\, dz \right| \to 0,
\tag{8.30}
$$

as $R \to \infty$. The device that makes this technique work is that, as $R \to \infty$,

$$
\log(x + i/\sqrt{2R}) - \log(x - i/\sqrt{2R}) \to -2\pi i.
\tag{8.31}
$$

So

$$
\oint_{\eta_R^1} g(z)\, dz + \oint_{\eta_R^3} g(z)\, dz \to -2\pi i \int_0^\infty \frac{dt}{t^2 + 6t + 8}.
\tag{8.32}
$$

Now (8.28)–(8.32) taken together yield

$$
\int_0^\infty \frac{dt}{t^2 + 6t + 8} = \frac{1}{2} \log 2 \,.
\tag{8.33}
$$

8.2.7 Summing a Series Using Residues

We sum the series

$$\sum_{j=1}^{\infty} \frac{x}{j\pi(j^2\pi^2 - x^2)} \tag{8.34}$$

using contour integration. Define $\cot z = \cos z / \sin z$. For $n = 1, 2, \ldots$ let Γ_n be the contour (shown in Figure 8.12) consisting of the counterclockwise oriented square with corners $\{(\pm 1 \pm i) \cdot (n + \frac{1}{2}) \cdot \pi\}$. For z fixed and $n > |z|$ we calculate using residues that

$$\frac{1}{2\pi i} \oint_{\Gamma_n} \frac{\cot \zeta}{\zeta(\zeta - z)} d\zeta = \sum_{j=1}^{n} \frac{1}{j\pi(j\pi - z)} + \sum_{j=1}^{n} \frac{1}{j\pi(j\pi + z)}$$

$$+ \frac{\cot z}{z} - \frac{1}{z^2}.$$

When $n \gg |z|$, it is easy to estimate the left-hand side in modulus by

$$\left(\frac{1}{2\pi}\right) \cdot [4(2n+1)\pi] \cdot \left(\frac{C}{n(n - |z|)}\right) \to 0 \quad \text{as} \quad n \to \infty. \tag{8.35}$$

Thus we see that

$$\sum_{j=1}^{\infty} \frac{1}{j\pi(j\pi - z)} + \sum_{j=1}^{\infty} \frac{1}{j\pi(j\pi + z)} = -\frac{\cot z}{z} + \frac{1}{z^2}. \tag{8.36}$$

We conclude that

$$\sum_{j=1}^{\infty} \frac{2}{j\pi(j^2\pi^2 - z^2)} = -\frac{\cot z}{z} + \frac{1}{z^2} \tag{8.37}$$

or

$$\sum_{j=1}^{\infty} \frac{z}{j\pi(j^2\pi^2 - z^2)} = -\frac{1}{2}\cot z + \frac{1}{2z}. \tag{8.38}$$

This is the desired result.

8.2.8 Summary Chart of Some Integration Techniques

In what follows we present, in chart form, just a few of the key methods of using residues to evaluate definite integrals.

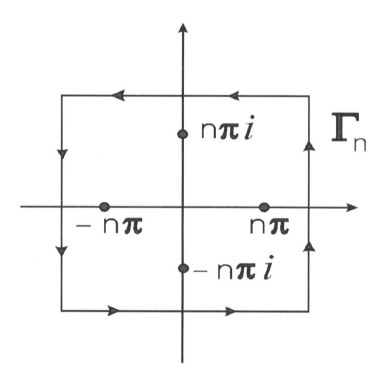

Figure 8.12: The curve Γ_n in Subsection 8.2.7.

Use of Residues to Evaluate Integrals

Integral	Properties of	Value of Integral
$I = \int_{-\infty}^{\infty} f(x)\,dx$	Finite number of poles of $f(z)$ in plane. $\lvert f(z) \rvert \leq \frac{C}{\lvert z \rvert^2}$ for z large.	$\begin{pmatrix} \text{sum of residues} \\ \text{of } f \text{ in upper} \\ \text{halfplane} \end{pmatrix}$
$I = \int_{-\infty}^{\infty} f(x)\,dx$	$f(z)$ may have simple poles on real axis. Finite number of poles of $f(z)$ in plane. $\lvert f(z) \rvert \leq \frac{C}{\lvert z \rvert^2}$ for z large.	$I = 2\pi i \times \begin{pmatrix} \text{sum of residues} \\ \text{of } f \text{ in upper} \\ \text{halfplane} \end{pmatrix}$ $+ \pi i \times \begin{pmatrix} \text{sum of residues} \\ \text{of } f(z) \\ \text{on real axis} \end{pmatrix}$
$I = \int_{-\infty}^{\infty} \frac{p(x)}{q(x)}\,dx$	p, q polynomials. $[\deg p] + 2 \leq \deg q$. q has no real zeros.	$I = 2\pi i \times \begin{pmatrix} \text{sum of residues} \\ \text{of } p(z)/q(z) \\ \text{in upper half} \\ \text{plane} \end{pmatrix}$

Use of Residues to Evaluate Integrals, Continued

Integral	Properties of	Value of Integral
$I =$ $$\int_{-\infty}^{\infty} \frac{p(x)}{q(x)}\, dx$$	p, q polynomials. $[\deg p] + 2 \leq \deg q$. $p(z)/q(z)$ may have simple poles on real axis.	$I = 2\pi i \times$ $\begin{pmatrix} \text{sum of residues} \\ \text{of } p(z)/q(z) \\ \text{in upper half} \\ \text{plane} \end{pmatrix}$ $+ \pi i \times$ $\begin{pmatrix} \text{sum of residues} \\ \text{of } p(z)/q(z) \\ \text{on real axis} \end{pmatrix}$
$I =$ $$\int_{-\infty}^{\infty} e^{i\alpha x} \cdot f(x)\, dx$$	$\alpha > 0$, z large $|f(z)| \leq \frac{C}{|z|}$ No poles of f on real axis.	$I = 2\pi i \times$ $\begin{pmatrix} \text{sum of residues} \\ \text{of } e^{i\alpha z} f(z) \\ \text{in upper half} \\ \text{plane} \end{pmatrix}$
$I =$ $$\int_{-\infty}^{\infty} e^{i\alpha x} \cdot f(x)\, dx$$	$\alpha > 0$, z large $|f(z)| \leq \frac{C}{|z|}$ $f(z)$ may have simple poles on real axis	$I = 2\pi i \times$ $\begin{pmatrix} \text{sum of residues} \\ \text{of } e^{i\alpha z} f(z) \\ \text{in upper half} \\ \text{plane} \end{pmatrix}$ $+ \pi i \times$ $\begin{pmatrix} \text{sum of residues} \\ \text{of } e^{i\alpha z} f(z) \\ \text{on real axis} \end{pmatrix}$

Exercises

Use the calculus of residues to calculate the integrals in Exercises 1–13:

1. $$\int_{0}^{+\infty} \frac{1}{1 + x^2}\, dx$$

2. $\displaystyle \int_{-\infty}^{+\infty} \frac{\cos x}{1+x^4}\, dx$

3. $\displaystyle \int_{0}^{+\infty} \frac{x^{1/4}}{1+x^3}\, dx$

4. $\displaystyle \int_{0}^{+\infty} \frac{1}{x^3+x+1}\, dx$

5. $\displaystyle \int_{0}^{+\infty} \frac{1}{1+x^3}\, dx$

6. $\displaystyle \int_{0}^{+\infty} \frac{x \sin x}{1+x^2}\, dx$

7. $\displaystyle \int_{0}^{\infty} \frac{1}{p(x)}\, dx$ where $p(x)$ is any polynomial with no zeros
on the nonnegative real axis

8. $\displaystyle \int_{-\infty}^{+\infty} \frac{x}{\sinh x}\, dx$

9. $\displaystyle \int_{-\infty}^{0} \frac{x^{1/3}}{-1+x^5}\, dx$

10. $\displaystyle \int_{-\infty}^{+\infty} \frac{\sin^2 x}{x^2}\, dx$

11. $\displaystyle \int_{-\infty}^{+\infty} \frac{x^4}{1+x^{10}}\, dx$

12. $\displaystyle \int_{-\infty}^{\infty} \frac{e^{-ix}}{\sqrt{x+2i}+\sqrt{x+5i}}\, dx$

13. $\displaystyle \int_{-\pi}^{\pi} \frac{d\theta}{5+3\cos\theta}$

Use the calculus of residues to sum each of the series in Exercises 14–
16. Make use of either the cotangent or tangent functions (as in the
text) to introduce infinitely many poles that are located at the integer
values that you wish to study.

14. $\displaystyle\sum_{k=0}^{\infty} \frac{1}{k^4 + 1}$

15. $\displaystyle\sum_{j=-\infty}^{\infty} \frac{1}{j^3 + 2}$

16. $\displaystyle\sum_{j=0}^{\infty} \frac{j^2 + 1}{j^4 + 4}$

17. When α is not a real integer, show that

$$\sum_{k=-\infty}^{\infty} \frac{1}{(k+\alpha)^2} = \frac{\pi^2}{\sin^2 \pi\alpha}.$$

Chapter 9

The Argument Principle

9.1 Counting Zeros and Poles

9.1.1 Local Geometric Behavior of a Holomorphic Function

In this chapter, we shall be concerned with questions that have a geometric, qualitative nature rather than an analytical, quantitative one. These questions center around the issue of the local geometric behavior of a holomorphic function.

9.1.2 Locating the Zeros of a Holomorphic Function

Suppose that $f : U \to \mathbb{C}$ is a holomorphic function on a connected, open set $U \subseteq \mathbb{C}$ and that $\overline{D}(P, r) \subseteq U$. We know from the Cauchy integral formula that the values of f on $D(P, r)$ are completely determined by the values of f on $\partial D(P, r)$. In particular, the number and even the location of the zeros of f in $D(P, r)$ are determined in principle by f on $\partial D(P, r)$. But it is nonetheless a pleasant surprise that there is a *simple formula* for the number of zeros of f in $D(P, r)$ in terms of f (and f') on $\partial D(P, r)$. In order to obtain a precise formula, we shall have to agree to count zeros according to multiplicity (see §§5.1.4). We now explain the precise idea.

Let $f : U \to \mathbb{C}$ be holomorphic as before, and assume that f has *some* zeros in U but that f is not identically zero. Fix $z_0 \in U$ such that

$f(z_0) = 0$. Since the zeros of f are isolated, there is an $r > 0$ such that $\overline{D}(z_0, r) \subseteq U$ and such that f does not vanish on $\overline{D}(z_0, r) \setminus \{z_0\}$.

Now the power series expansion of f about z_0 has a first non-zero term determined by the least positive integer n such that $f^{(n)}(z_0) \neq 0$. (Note that $n \geq 1$ since $f(z_0) = 0$ by hypothesis.) Thus the power series expansion of f about z_0 *begins* with the nth term:

$$f(z) = \sum_{j=n}^{\infty} \frac{1}{j!} \frac{\partial^j f}{\partial z^j}(z_0)(z - z_0)^j. \tag{9.1}$$

Under these circumstances we say that f has a zero of *order n* (or *multiplicity n*) at z_0. When $n = 1$, then we also say that z_0 is a *simple* zero of f.

The important point to see here is that, near z_0,

$$\frac{f'(z)}{f(z)} \approx \frac{[n/n!] \cdot (\partial^n f/\partial z^n)(z_0)(z - z_0)^{n-1}}{[1/n!] \cdot (\partial^n f/\partial z^n)(z_0)(z - z_0)^n} = \frac{n}{z - z_0}.$$

It follows then that

$$\frac{1}{2\pi i} \oint_{\partial D(z_0, r)} \frac{f'(z)}{f(z)} \, dz \approx \frac{1}{2\pi i} \oint_{\partial D(z_0, r)} \frac{n}{z - z_0} \, dz = n \, .$$

On the one hand, this is an approximation. On the other hand, the approximation becomes more and more true as r shrinks to 0. And the value of the integral is independent of r. Thus we may conclude that we have equality. Amazingly, the value of the integral is an integer.

In short, the complex line integral of f'/f around the boundary of the disc gives the order of the zero at the center. If there are several zeros of f inside the disc $D(z_0, r)$ then we may break the complex line integral up into individual integrals around each of the zeros (see Figure 9.1), so we have the more general result that the integral of f'/f counts *all* the zeros inside the disc, together with their multiplicities. We shall consider this idea further in the discussion that follows.

9.1.3   Zero of Order n

The concept of zero of "order n," or "multiplicity n," for a function f is so important that a variety of terminology has grown up around it (see

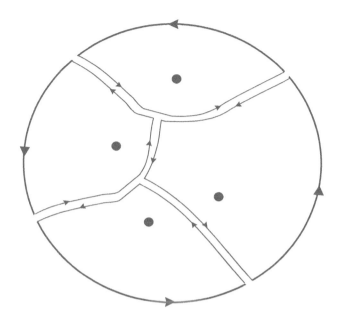

Figure 9.1: Dividing up the complex line integral.

also §§5.1.4). It has already been noted that when the multiplicity $n = 1$, then the zero is sometimes called *simple*. For arbitrary n, we sometimes say that "n is the order of z_0 as a zero of f." More generally if $f(z_0) = \beta$ so that, for some $n \geq 1$, the function $f(\cdot) - \beta$ has a zero of order n at z_0, then we say either that "f assumes the value β at z_0 to order n" or that "the order of the value β at z_0 is n." When $n > 1$, then we call z_0 a *multiple point* of the function f.

The next result provides a method for computing the multiplicity n of the zero at z_0 from the values of f, f' on the boundary of a disc centered at z_0.

9.1.4 Counting the Zeros of a Holomorphic Function

If f is holomorphic on a neighborhood of a disc $\overline{D}(P, r)$ and has a zero of order n at P and no other zeros in the closed disc, then

$$\frac{1}{2\pi i} \oint_{\partial D(P,r)} \frac{f'(\zeta)}{f(\zeta)} \, d\zeta = n. \tag{9.2}$$

More generally, we consider the case that f has several zeros—with different locations and different multiplicities—inside a disc: Suppose that $f : U \to \mathbb{C}$ is holomorphic on an open set $U \subseteq \mathbb{C}$ and that $\overline{D}(P, r) \subseteq U$. Suppose that f is nonvanishing on $\partial D(P, r)$ and that $z_1, z_2, \ldots, z_k$ are the zeros of f in the interior of the disc. Let n_ℓ be the order of the zero of f at z_ℓ, $\ell = 1, \ldots, k$. Then

$$\frac{1}{2\pi i} \oint_{|\zeta - P| = r} \frac{f'(\zeta)}{f(\zeta)} \, d\zeta = \sum_{\ell=1}^{k} n_\ell. \tag{9.3}$$

Refer to Figure 9.2 for illustrations of both these situations.

It is worth noting that the particular features of a *circle* play no particular role in these considerations. We could as well consider the zeros of a function f that lie inside a simple, closed curve γ. Then it still holds that

$$(\text{number of zeros, counting multiplicity}) = \frac{1}{2\pi i} \oint_{\gamma} \frac{f'(z)}{f(z)} \, dz \, .$$

9.1.5   The Argument Principle

This last formula, which is often called the *argument principle*, is both useful and important. For one thing, there is no obvious reason why the integral in the formula should be an integer, much less the crucial integer that it is. Since it is an integer, it is a counting function; and we need to learn more about it.

The integral

$$\frac{1}{2\pi i} \oint_{|\zeta - P| = r} \frac{f'(\zeta)}{f(\zeta)} \, d\zeta \tag{9.4}$$

can be reinterpreted as follows: Consider the C^1 closed curve

$$\gamma(t) = f(P + re^{it}), \quad t \in [0, 2\pi]. \tag{9.5}$$

Then

$$\frac{1}{2\pi i} \oint_{|\zeta - P| = r} \frac{f'(\zeta)}{f(\zeta)} \, d\zeta = \frac{1}{2\pi i} \int_{0}^{2\pi} \frac{\gamma'(t)}{\gamma(t)} \, dt, \tag{9.6}$$

as you can check by direct calculation. The expression on the right is just the index of the curve γ with respect to 0 (with the notion of index that

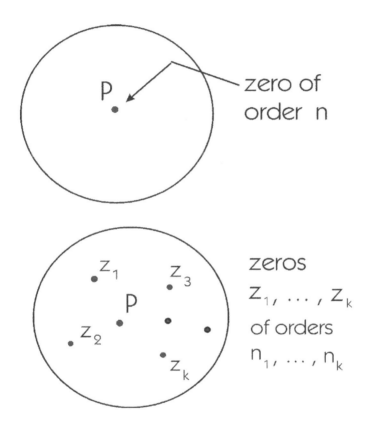

Figure 9.2: Locating the zeros of a holomorphic function.

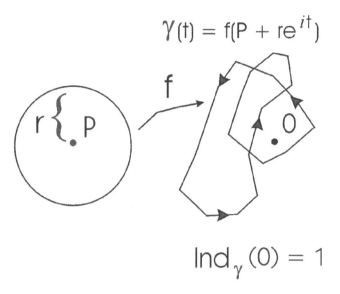

$$\gamma(t) = f(P + re^{it})$$

$$\text{Ind}_\gamma(0) = 1$$

Figure 9.3: The argument principle: counting the zeros.

we defined earlier—§§8.1.5). See Figure 9.3. Thus the number of zeros of f (counting multiplicity) inside the circle $\{\zeta : |\zeta - P| = r\}$ is equal to the index of γ with respect to the origin. This, intuitively speaking, is equal to the number of times that the f-image of the boundary circle winds around 0 in $\mathbb{C}$. So we have another way of seeing that the value of the integral must be an integer.

The argument principle can be extended to yield information about meromorphic functions, too. We can see that there is hope for this notion by investigating the analog of the argument principle for a pole.

9.1.6   Location of Poles

If $f : U \setminus \{Q\} \to \mathbb{C}$ is a nowhere-zero holomorphic function on $U \setminus \{Q\}$ with a pole of order n at Q and if $\overline{D}(Q,r) \subseteq U$, then

$$\frac{1}{2\pi i} \oint_{\partial D(Q,r)} \frac{f'(\zeta)}{f(\zeta)} \, d\zeta = -n. \tag{9.7}$$

The argument is just the same as the calculations we did right after formula (9.1) (or else think about the fact that if f has a pole of order k at Q then $1/f$ has a zero of order k at Q). We shall not repeat the details, but we invite the reader to do so.

9.1.7 The Argument Principle for Meromorphic Functions

Just as with the argument principle for holomorphic functions, this new argument principle gives a counting principle for zeros and poles of meromorphic functions:

Suppose that f is a meromorphic function on an open set $U \subseteq \mathbb{C}$, that $\overline{D}(P,r) \subseteq U$, and that f has neither poles nor zeros on $\partial D(P,r)$. Then

$$\frac{1}{2\pi i} \oint_{\partial D(P,r)} \frac{f'(\zeta)}{f(\zeta)} \, d\zeta = \sum_{j=1}^{p} n_j - \sum_{k=1}^{q} m_k, \qquad (9.8)$$

where $n_1, n_2, \ldots, n_p$ are the multiplicities of the zeros $z_1, z_2, \ldots, z_p$ of f in $D(P,r)$ and $m_1, m_2, \ldots, m_q$ are the orders of the poles $w_1, w_2, \ldots, w_q$ of f in $D(P,r)$.

Of course the reasoning here is by now familiar. We can break up the complex line integral around the boundary of the disc $D(P,r)$ into integrals around smaller regions, each of which contains just one zero or one pole and no other. Refer again to Figure 9.1. Thus the integral around the disc just sums up $+k$ for each zero of order k and $-m$ for each pole of order m.

Exercises

1. Use argument principle to give another verification of the Fundamental Theorem of Algebra. [**Hint:** Think about the integral of $p'(z)/p(z)$ over circles centered at the origin of larger and larger radius.]

2. Imitate the verification of the argument principle to demonstrate the following formula: If $f : U \to \mathbb{C}$ is holomorphic in U and invertible, $P \in U$, and if $D(P,r)$ is a sufficiently small disc about P, then

$$f^{-1}(w) = \frac{1}{2\pi i} \oint_{\partial D(P,r)} \frac{\zeta f'(\zeta)}{f(\zeta) - w} \, d\zeta$$

for all w in some disc $D(f(P), r_1)$, $r_1 > 0$ sufficiently small. Derive from this the formula

$$(f^{-1})'(w) = \frac{1}{2\pi i} \oint_{\partial D(P,r)} \frac{\zeta f'(\zeta)}{(f(\zeta) - w)^2} \, d\zeta.$$

Set $Q = f(P)$. Integrate by parts and use some algebra to obtain

$$(f^{-1})'(w) = \frac{1}{2\pi i} \oint_{\partial D(P,r)} \left(\frac{1}{f(\zeta) - Q} \right) \cdot \left(1 - \frac{w - Q}{f(\zeta) - Q} \right)^{-1} d\zeta. \quad (9.9)$$

Let a_k be the kth coefficient of the power series expansion of f^{-1} about the point Q :

$$f^{-1}(w) = \sum_{k=0}^{\infty} a_k (w - Q)^k.$$

Then formula (9.9) may be expanded and integrated term by term (demonstrate this!) to obtain

$$\begin{aligned} na_n &= \frac{1}{2\pi i} \oint_{\partial D(P,r)} \frac{1}{[f(\zeta) - Q]^n} d\zeta \\ &= \frac{1}{(n-1)!} \left(\frac{\partial}{\partial \zeta} \right)^{n-1} \frac{(\zeta - P)^n}{[f(\zeta) - Q]^n} \bigg|_{\zeta=P}. \end{aligned}$$

This is called *Lagrange's formula*.

3. Suppose that f is holomorphic and has n zeros, counting multiplicities, inside U. Can you conclude that f' has $(n-1)$ zeros inside U? Can you conclude anything about the zeros of f'?

4. **Demonstrate:** If f is a polynomial on $\mathbb{C}$, then the zeros of f' are contained in the closed convex hull of the zeros of f. (Here the *closed convex hull* of a set S is the intersection of all closed convex sets that contain S.) [**Hint:** If the zeros of f are contained in a halfplane V, then so are the zeros of f'.]

5. Let $P_t(z)$ be a polynomial in z for each fixed value of $t, 0 \le t \le 1$. Suppose that $P_t(z)$ is continuous in t in the sense that

$$P_t(z) = \sum_{j=0}^{N} a_j(t) z^j$$

and each $a_j(t)$ is continuous. Let $\mathcal{Z} = \{(z, t) : P_t(z) = 0\}$. By continuity, $\mathcal{Z}$ is closed in $\mathbb{C} \times [0, 1]$. If $P_{t_0}(z_0) = 0$ and $(\partial/\partial z) P_{t_0}(z) \big|_{z=z_0} \ne 0$, then show, using the argument principle, that there is an $\epsilon > 0$ such that for t sufficiently near t_0 there is a unique $z \in D(z_0, \epsilon)$ with $P_t(z) = 0$. What can you say if $P_{t_0}(\cdot)$ vanishes to order k at z_0?

6. Show that if $f : U \to \mathbb{C}$ is holomorphic, $P \in U$, and $f'(P) = 0$, then f is not one-to-one in any neighborhood of P.

7. **Demonstrate:** If f is holomorphic on a neighborhood of the closed unit disc D and if f is one-to-one on ∂D, then f is one-to-one on $\overline{D}$. [*Note:* Here you may assume any topological notions you need that seem intuitively plausible. Remark on each one as you use it.]

8. Let $p_t(z) = a_0(t) + a_1(t)z + \cdots + a_n(t)z^n$ be a polynomial in which the coefficients depend continuously on a parameter $t \in (-1, 1)$. Show that if the roots of p_{t_0} are distinct (no multiple roots), for some fixed value of the parameter, then the same is true for p_t when t is sufficiently close to t_0—*provided* that the degree of p_t remains the same as the degree of p_{t_0}.

9.2 The Local Geometry of Holomorphic Functions

9.2.1 The Open Mapping Theorem

The argument principle for holomorphic functions has a consequence that is one of the most important facts about holomorphic functions considered as geometric mappings:

THEOREM 1 *If $f : U \to \mathbb{C}$ is a non-constant holomorphic function on a connected open set U, then $f(U)$ is an open set in $\mathbb{C}$.*

See Figure 9.4. The result says, in particular, that if $U \subseteq \mathbb{C}$ is connected and open and if $f : U \to \mathbb{C}$ is holomorphic, then either $f(U)$ is a connected open set (the non-constant case) or $f(U)$ is a single point.

The open mapping principle has some interesting and important consequences. Among them are:

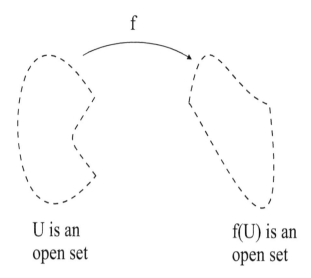

f

U is an
open set

f(U) is an
open set

Figure 9.4: The open mapping principle.

(a) If U is a domain in $\mathbb{C}$ and $f : U \to \mathbb{R}$ is a holomorphic function then
f must be constant. For the theorem says that the image of f must
be *open* (as a subset of the plane), and the real line contains no planar
open sets.

(b) Let U be a domain in $\mathbb{C}$ and $f : U \to \mathbb{C}$ a holomorphic function.
Suppose that the set E lies in the image of f. Then the image of f
must in fact contain a neighborhood of E.

(c) Let U be a domain in $\mathbb{C}$ and $f : U \to \mathbb{C}$ a holomorphic function. Let
$P \in U$ and set $k = |f(P)|$. Then k cannot be the maximum value of
$|f|$. For in fact (by part (b)) the image of f must contain an entire
neighborhood of $f(P)$. So (see Figure 9.5), it will certainly contain
points with modulus larger than k. This is a version of the important
maximum principle which we shall discuss in some detail below.

In fact the open mapping principle is an immediate consequence of
the argument principle. For suppose that $f : U \to \mathbb{C}$ is holomorphic
and that $P \in U$. Write $f(P) = Q$. We may select an $r > 0$ so that
$\overline{D}(P, r) \subseteq U$. Let $g(z) = f(z) - Q$. Then g has a zero at P.

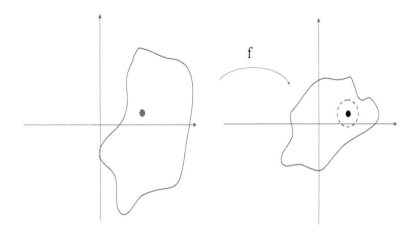

Figure 9.5: The maximum principle follows from the open mapping principle.

The argument principle now tells us that

$$\frac{1}{2\pi i} \oint_{\partial D(P,r)} \frac{g'(z)}{g(z)}\, dz \geq 1\,.$$

[We do not write $= 1$ because we do not know the order of vanishing of g—but it is *at least* 1.] In other words,

$$\frac{1}{2\pi i} \oint_{\partial D(P,r)} \frac{f'(z)}{f(z) - Q}\, dz \geq 1\,.$$

But now the continuity of the integral tells us that, if we perturb Q by a small amount, then the value of the integral—which still must be an integer!—will not change. This says that f assumes values that are near to Q. Which says that the image of f is open. That is the assertion of the open mapping principle.

In the subject of topology, a function f is defined to be continuous if the inverse image of any open set under f is also open. In contexts where the $\epsilon - \delta$ definition makes sense, the $\epsilon - \delta$ definition (§§3.1.6) is equivalent to the inverse-image-of-open-sets definition. By contrast, functions for which the direct image of any open set is open are called "open mappings."

Here is a quantitative, or counting, statement that comes from the verification of the open mapping principle: Suppose that $f : U \to \mathbb{C}$

is a non-constant holomorphic function on a connected open set U such that $P \in U$ and $f(P) = Q$ with order $k \geq 2$. Then there are numbers $\delta, \epsilon > 0$ such that each $q \in D(Q, \epsilon) \setminus \{Q\}$ has exactly k distinct pre-images in $D(P, \delta)$ and each pre-image is a simple point of f. This is a striking statement; but all we are saying is that the set of points where f' vanishes cannot have an interior accumulation point.

The considerations that establish the open mapping principle can also be used to establish the fact that if $f : U \to V$ is a one-to-one and onto holomorphic function, then $f^{-1} : V \to U$ is also holomorphic.

Exercises

1. Let f be holomorphic on a neighborhood of $\overline{D}(P, r)$. Suppose that f is not identically zero on $D(P, r)$. Show that f has at most finitely many zeros in $D(P, r)$.

2. Let f, g be holomorphic on a neighborhood $\overline{D}(0, 1)$. Assume that f has zeros at $P_1, P_2, \ldots, P_k \in D(0, 1)$ and no zero in $\partial D(0, 1)$. Let γ be the boundary circle of $\overline{D}(0, 1)$, traversed counterclockwise. Compute

    $$\frac{1}{2\pi i} \oint_\gamma \frac{f'(z)}{f(z)} \cdot g(z) dz.$$

3. Without supposing that you have any prior knowledge of the calculus function e^x, show that

    $$e^z \equiv \sum_{k=0}^{\infty} \frac{z^k}{k!}$$

    never vanishes by computing $(e^z)'/e^z$, and so forth.

4. Let $f_j : D(0, 1) \to \mathbb{C}$ be holomorphic and suppose that each f_j has at least k roots in $D(0, 1)$, counting multiplicities. Suppose that $f_j \to f$ uniformly on compact sets. Show by example that it does *not* follow that f has at least k roots counting multiplicities. In particular, construct examples, for each fixed k and each ℓ, $0 \leq \ell \leq k$, where f has exactly ℓ roots. What simple hypothesis can you add that will guarantee that f *does* have at least k roots?

5. Let $f : D(0,1) \to \mathbb{C}$ be holomorphic and nonvanishing. Show that f has a well-defined holomorphic logarithm on $D(0,1)$ by showing that the differential equation

$$\frac{\partial}{\partial z} g(z) = \frac{f'(z)}{f(z)}$$

has a suitable solution and checking that this solution g does the job.

6. Let U and V be open subsets of $\mathbb{C}$. Suppose that $f : U \to V$ is holomorphic, one-to-one, and onto. Show that f^{-1} is a holomorphic function on V.

7. Let $f : U \to \mathbb{C}$ be holomorphic. Assume that $\overline{D}(P,r) \subseteq U$ and that f is nowhere zero on $\partial D(P,r)$. Show that if g is holomorphic on U and g is sufficiently uniformly close to f on $\partial D(P,r)$, then the number of zeros of f in $D(P,r)$ equals the number of zeros of g in $D(P,r)$. (Remember to count zeros according to multiplicity.)

8. Estimate the number of zeros of the given function in the given region U.

(a) $f(z) = z^8 + 5z^7 - 20$,	$U = D(0,6)$
(b) $f(z) = z^3 - 3z^2 + 2$,	$U = D(0,1)$
(c) $f(z) = z^{10} + 10z + 9$,	$U = D(0,1)$
(d) $f(z) = z^{10} + 10ze^{z+1} - 9$,	$U = D(0,1)$
(e) $f(z) = z^4 e - z^3 + z^2/6 - 10$,	$U = D(0,2)$
(f) $f(z) = z^2 e^z - z$,	$U = D(0,2)$

9. Imitate the verification of the argument principle to demonstrate the following formula: If $f : U \to \mathbb{C}$ is holomorphic in U and invertible, $P \in U$, and if $D(P,r)$ is a sufficiently small disc about P, then

$$f^{-1}(w) = \frac{1}{2\pi i} \oint_{\partial D(P,r)} \frac{\zeta f'(\zeta)}{f(\zeta) - w} \, d\zeta$$

for all w in some disc $D(f(P), r_1)$, $r_1 > 0$ sufficiently small. Derive from this the formula

$$(f^{-1})'(w) = \frac{1}{2\pi i} \oint_{\partial D(P,r)} \frac{\zeta f'(\zeta)}{(f(\zeta) - w)^2} \, d\zeta.$$

Set $Q = f(P)$. Integrate by parts and use some algebra to obtain

$$(f^{-1})'(w) = \frac{1}{2\pi i} \oint_{\partial D(P,r)} \left(\frac{1}{f(\zeta) - Q} \right) \cdot \left(1 - \frac{w - Q}{f(\zeta) - Q} \right)^{-1} d\zeta. \quad (9.10)$$

Let a_k be the kth coefficient of the power series expansion of f^{-1} about the point Q :

$$f^{-1}(w) = \sum_{k=0}^{\infty} a_k (w - Q)^k.$$

Then formula (9.10) may be expanded and integrated term by term (demonstrate this!) to obtain

$$
\begin{aligned}
n a_n &= \frac{1}{2\pi i} \oint_{\partial D(P,r)} \frac{1}{[f(\zeta) - Q]^n} d\zeta \\
&= \frac{1}{(n-1)!} \left(\frac{\partial}{\partial \zeta} \right)^{n-1} \frac{(\zeta - P)^n}{[f(\zeta) - Q]^n} \bigg|_{\zeta = P}.
\end{aligned}
$$

This is called *Lagrange's formula.*

9.3 Further Results on the Zeros of Holomorphic Functions

9.3.1 Rouché's Theorem

Now we consider global aspects of the argument principle.

Suppose that $f, g : U \to \mathbb{C}$ are holomorphic functions on an open set $U \subseteq \mathbb{C}$. Suppose also that $\overline{D}(P,r) \subseteq U$ and that, for each $\zeta \in \partial D(P,r)$,

$$|f(\zeta) - g(\zeta)| < |f(\zeta)| + |g(\zeta)|. \quad (9.11)$$

Then

$$\frac{1}{2\pi i} \oint_{\partial D(P,r)} \frac{f'(\zeta)}{f(\zeta)} d\zeta = \frac{1}{2\pi i} \oint_{\partial D(P,r)} \frac{g'(\zeta)}{g(\zeta)} d\zeta. \quad (9.12)$$

That is, the number of zeros of f in $D(P,r)$ counting multiplicities equals the number of zeros of g in $D(P,r)$ counting multiplicities. This result

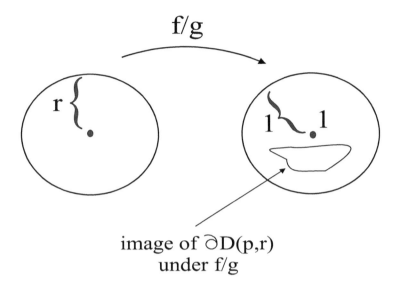

f/g

image of $\partial D(p,r)$
under f/g

Figure 9.6: Rouché's theorem.

is called *Rouché's theorem.*

Remark: Rouché's theorem is often stated with the stronger hypothesis that

$$|f(\zeta) - g(\zeta)| < |g(\zeta)| \qquad (9.13)$$

for $\zeta \in \partial D(P, r)$. Rewriting this hypothesis as

$$\left| \frac{f(\zeta)}{g(\zeta)} - 1 \right| < 1, \qquad (9.14)$$

we see that it says that the image γ under f/g of the circle $\partial D(P, r)$ lies in the disc $D(1, 1)$. See Figure 9.6. Our weaker hypothesis that $|f(\zeta) - g(\zeta)| < |f(\zeta)| + |g(\zeta)|$ has the geometric interpretation that $f(\zeta)/g(\zeta)$ lies in the set $\mathbb{C} \setminus \{x + i0 : x \leq 0\}$. Either hypothesis implies that the image of the circle $\partial D(P, r)$ under f has the same "winding number" around 0 as does the image under g of that circle. And that is the verification of Rouché's theorem.

9.3.2 A Typical Application of Rouché's Theorem

EXAMPLE 7 Let us determine the number of roots of the polynomial $f(z) = z^7 + 5z^3 - z - 2$ in the unit disc. We do so by comparing the function f to the holomorphic function $g(z) = 5z^3$ on the unit circle. For $|z| = 1$ we have

$$|f(z) - g(z)| = |z^7 - z - 2| \le 4 < |g(\zeta)| \le |f(\zeta)| + |g(\zeta)|. \qquad (9.15)$$

By Rouché's theorem, f and g have the same number of zeros, counting multiplicity, in the unit disc. Since g has three zeros, so does f. □

9.3.3 Rouché's Theorem and the Fundamental Theorem of Algebra

Rouché's theorem provides a useful way to locate approximately the zeros of a holomorphic function that is too complicated for the zeros to be obtained explicitly. As an illustration, we analyze the zeros of a non-constant polynomial

$$P(z) = z^n + a_{n-1}z^{n-1} + a_{n-2}z^{n-2} + \cdots + a_1 z + a_0. \qquad (9.16)$$

If R is sufficiently large (say $R > \max\{1, n \cdot \max_{0 \le j \le n-1} |a_j|\}$) and $|z| = R$, then

$$\frac{|a_{n-1}z^{n-1} + a_{n-2}z^{n-2} + \cdots + a_0|}{|z^n|} < 1. \qquad (9.17)$$

Thus Rouché's theorem applies on $\overline{D}(0, R)$ with $f(z) = z^n$ and $g(z) = P(z)$. We conclude that the number of zeros of $P(z)$ inside $D(0, R)$, counting multiplicities, is the same as the number of zeros of z^n inside $D(0, R)$, counting multiplicities—namely n. Thus we recover the Fundamental Theorem of Algebra. Incidentally, this example underlines the importance of counting zeros with multiplicities: the function z^n has only one root in the naïve sense of counting the number of points where it is zero; but it has n roots when they are counted with multiplicity.

9.3.4 Hurwitz's Theorem

A second useful consequence of the argument principle is the following result about the limit of a sequence of zero-free holomorphic functions:

THEOREM 2 (Hurwitz's Theorem) *Suppose that $U \subseteq \mathbb{C}$ is a connected open set and that $\{f_j\}$ is a sequence of nowhere-vanishing holomorphic functions on U. If the sequence $\{f_j\}$ converges uniformly on compact subsets of U to a (necessarily holomorphic) limit function f_0, then either f_0 is nowhere-vanishing or $f_0 \equiv 0$.*

The justification for Hurwitz's theorem is again the argument principle. For we know that if $\overline{D}(P, r)$ is a closed disc on which all the f_j are zero-free then

$$\frac{1}{2\pi i} \oint_{\partial D(P,r)} \frac{f_j'(z)}{f_j(z)}\, dz = 0$$

for every j. Since the f_j (and hence the f_j') converge uniformly on $\overline{D}(P, r)$, we can be sure that as $j \to +\infty$ the expression on the left converges to

$$\frac{1}{2\pi i} \oint_{\partial D(P,r)} \frac{f'(z)}{f(z)}\, dz .$$

We conclude that f has no zeros in the disc either—unless $f \equiv 0$, in which case this limiting argument is invalid.

Exercises

1. How many zeros does the function $f(z) = z^3 + z/2$ have in the unit disc?

2. Consider the sequence of functions $f_j(z) = e^{z/j}$. Discuss this sequence in view of Hurwitz's theorem.

3. Consider the sequence of functions $f_j(z) = \sin(jz)$. Discuss in view of Hurwitz's theorem.

4. Consider the sequence of functions $f_j(z) = \cos(z/j)$. Discuss in view of Hurwitz's theorem.

5. Apply Rouché's theorem to see that e^z cannot vanish on the unit disc.

6. Each of the partial sums of the power series for the function e^z is a polynomial. Hence it has zeros. But the exponential function has no zeros. Discuss in view of Hurwitz's theorem and the argument principle.

7.　Each of the partial sums of the power series for the function $\sin z$ is a polynomial, hence it has finitely many zeros. Yet $\sin z$ has infinitely many zeros. Discuss in view of Hurwitz's theorem and the argument principle.

8.　How many zeros does $f(z) = \sin z + \cos z$ have in the unit disc?

9.　Use the open mapping theorem to verify the maximum principle for *harmonic* functions. [**Hint:** If u is harmonic on a simply connected domain Ω, then let v be a harmonic conjugate of u. Then $h \equiv e^{u+iv}$ is holomorphic. Apply the open mapping theorem to h.]

Chapter 10

The Maximum Principle

10.1 Local and Boundary Maxima

10.1.1 The Maximum Modulus Principle

A *domain* in $\mathbb{C}$ is a connected open set (§§1.2.2). A *bounded domain* is a connected open set U such that there is an $R > 0$ with $|z| < R$ for all $z \in U$—or $U \subseteq D(0, R)$.

The Maximum Modulus Principle

> Let $U \subseteq \mathbb{C}$ be a domain. Let f be a holomorphic function on U. If there is a point $P \in U$ such that $|f(P)| \geq |f(z)|$ for all $z \in U$, then f is constant.

Here is a sharper variant of the theorem:

> Let $U \subseteq \mathbb{C}$ be a domain and let f be a holomorphic function on U. If there is a point $P \in U$ at which $|f|$ has a *local maximum*, then f is constant.

We have already indicated why this result is true; the geometric insight is an important one. Let $k = |f(P)|$. Since $f(P)$ is an *interior point* of the image of f, there will certainly be points—and the verification of the open mapping principle shows that these are nearby points—where f takes values of greater modulus. Hence P cannot be a local maximum.

10.1.2   Boundary Maximum Modulus Theorem

The following version of the maximum principle is intuitively appealing, and is frequently useful.

> Let $U \subseteq \mathbb{C}$ be a bounded domain. Let f be a continuous function on $\overline{U}$ that is holomorphic on U. Then the maximum value of $|f|$ on $\overline{U}$ (which must occur, since $\overline{U}$ is closed and bounded—see [RUD1], [KRA2]) must in fact occur on ∂U.
>
> In other words,
>
> $$\max_{\overline{U}} |f| = \max_{\partial U} |f|. \qquad (10.1)$$

And the reason for this new assertion is obvious. Since the domain is bounded, the maximum must occur somewhere; and it cannot occur in the interior by the previous formulation of the maximum principle. So it must be in the boundary.

10.1.3   The Minimum Principle

Holomorphic functions (or, more precisely, their moduli) *can* have interior minima. The function $f(z) = z^2$ on $D(0,1)$ has the property that $z = 0$ is a global minimum for $|f|$. However, it is not accidental that this minimum value is 0:

> Let f be holomorphic on a domain $U \subseteq \mathbb{C}$. Assume that f never vanishes. If there is a point $P \in U$ such that $|f(P)| \leq |f(z)|$ for all $z \in U$, then f is constant. This result is demonstrated by applying the maximum principle to the function $1/f$.

There is also a boundary minimum principle:

> Let $U \subseteq \mathbb{C}$ be a bounded domain. Let f be a continuous function on $\overline{U}$ that is holomorphic on U. Assume that f never vanishes on $\overline{U}$. Then the minimum value of $|f|$ on $\overline{U}$ (which must occur, since $\overline{U}$ is closed and bounded—see [RUD1], [KRA2])—must occur on ∂U.
>
> In other words,
>
> $$\min_{\overline{U}} |f| = \min_{\partial U} |f|. \qquad (10.2)$$

10.1.4 The Maximum Principle on an Unbounded Domain

It should be noted that the maximum modulus theorem is not always true on an unbounded domain. The standard example is the function $f(z) = \exp(\exp(z))$ on the domain $U = \{z = x + iy : -\pi/2 < y < \pi/2\}$. Check for yourself that $|f| = 1$ on the boundary of U. But the restriction of f to the real number line is unbounded at infinity. The theorem does, however, remain true with some additional restrictions. The result known as the Phragmen—Lindelöf theorem—which puts some additional hypotheses on the growth of the function at ∞—is one method of treating maximum modulus theorems on unbounded domains (see [RUD2]).

Exercises

1. Let $U \subseteq \mathbb{C}$ be a bounded domain. If f, g are continuous functions on $\overline{U}$, holomorphic on U, and if $|f(z)| \le |g(z)|$ for $z \in \partial U$, then what conclusion can you draw about f and g in the interior of U?

2. Let $f : \overline{D}(0,1) \to \overline{D}(0,1)$ be continuous and holomorphic on the interior. Further assume that f is one-to-one and onto. Explain why the maximum principle guarantees that $f(\partial D(0,1)) \subseteq \partial D(0,1)$.

3. Give an example of a holomorphic function f on $D(0,1)$ so that $|f|$ has three local minima.

4. Give an example of a holomorphic function f on $D(0,1)$, continuous on $\overline{D}(0,1)$, that has precisely three global maxima on $\partial D(0,1)$.

5. The function
$$f(z) = i \cdot \frac{1-z}{1+z}$$
maps the disc $D(0,1)$ to the upper halfplane $U = \{z \in \mathbb{C} : \operatorname{Im} z > 0\}$ (the upper halfplane) in a one-to-one, onto fashion. Verify this assertion in the following manner:

   (a) Use elementary algebra to check that f is one-to-one.

   (b) Use just algebra to check that $\partial D(0,1)$ is mapped to ∂U.

(c) Check that 0 is mapped to i.

(d) Use algebra to check that f is onto.

(e) Invoke the maximum principle to conclude that $D(0, 1)$ is mapped to U.

6.   Let f be meromorphic on a region $U \subseteq \mathbb{C}$. A version of the maximum principle is still valid for such an f. Explain why.

7.   Let $U \subseteq \mathbb{C}$ be a domain and let $f : U \to \mathbb{C}$ be holomorphic. Consider the function $g(z) = e^{f(z)}$. Explain why the maxima of $|g|$ occur precisely at the maxima of $\operatorname{Re} f$. Conclude that a version of the maximum principle holds for $\operatorname{Re} f$. Draw a similar conclusion for $\operatorname{Im} f$.

8.   Let U and V be domains in the plane. We call a function $f : U \to V$ *proper* if the image of any compact set is compact. Show that properness implies that if $\{z_j\} \subseteq U$ satisfy $z_j \to \partial U$, then $f(z_j) \to \partial V$. What does this have to do with the maximum principle?

10.2   The Schwarz Lemma

This section treats certain estimates that must be satisfied by bounded holomorphic functions on the unit disc. We present the classical, analytic viewpoint in the subject (instead of the geometric viewpoint—see [KRA3]).

10.2.1   Schwarz's Lemma

THEOREM 3 *Let f be holomorphic on the unit disc. Assume that*

(10.18) $|f(z)| \leq 1$ *for all* z.

(10.19) $f(0) = 0$.

Then $|f(z)| \leq |z|$ and $|f'(0)| \leq 1$.
If either $|f(z)| = |z|$ for some $z \neq 0$ or if $|f'(0)| = 1$, then f is a rotation: $f(z) \equiv \alpha z$ for some complex constant α of unit modulus.

To understand this result, consider the function $g(z) = f(z)/z$. Since g has a removable singularity at the origin, we see that g is holomorphic on the entire unit disc. On the circle with center 0 and radius $1 - \epsilon$, we see that

$$|g(z)| \leq \frac{1}{1 - \epsilon}.$$

By the maximum modulus principle, it follows that $|g(z)| \leq 1/(1 - \epsilon)$ on all of $\overline{D}(0, 1 - \epsilon)$. Since the conclusion is true for all $\epsilon > 0$, we conclude that $|g| \leq 1$ on $D(0, 1)$. Hence $|f(z)| \leq |z|$ for all z in the disc.

For the uniqueness, assume that $|f(z)| = |z|$ for some $z \neq 0$. Then $|g(z)| = 1$. Since $|g| \leq 1$ globally, the maximum modulus principle tells us that g is a constant of modulus 1. Thus $f(z) = \alpha z$ for some unimodular constant α. In conclusion, f is a rotation.

If instead $|f'(0)| = 1$, then $|[g(0) + g'(0) \cdot 0| = 1$ or $|g(0)| = 1$. Again, the maximum principle tells us that g is a unimodular constant, so f is a rotation.

Schwarz's lemma enables one to classify the invertible holomorphic self-maps of the unit disc (see [GRK]). (Here a *self-map* of a domain U is a mapping $F : U \rightarrow U$ of the domain to itself.) These are commonly referred to as the "conformal self-maps" of the disc. The classification is as follows: If $0 \leq \theta < 2\pi$, then define the *rotation through angle* θ to be the function $\rho_\theta(z) = e^{i\theta} z$; if a is a complex number of modulus less than one, then define the associated *Möbius transformation* to be $\varphi_a(z) = [z - a]/[1 - \bar{a}z]$. Any conformal self-map of the disc is the composition of some rotation ρ_θ with some Möbius transformation φ_a. This topic is treated in detail in §11.2.

We conclude this section by presenting a generalization of the Schwarz lemma, in which we consider holomorphic mappings $f : D \rightarrow D$, but we discard the hypothesis that $f(0) = 0$. This result is known as the Schwarz–Pick lemma.

10.2.2   The Schwarz–Pick Lemma

Let f be holomorphic on the unit disc. Assume that

6.20 $|f(z)| \leq 1$ for all z.

6.21 $f(a) = b$ for some $a, b \in D(0, 1)$.

Then

$$|f'(a)| \leq \frac{1 - |b|^2}{1 - |a|^2}. \tag{10.3}$$

Moreover, if $f(a_1) = b_1$ and $f(a_2) = b_2$, then

$$\left| \frac{b_2 - b_1}{1 - \bar{b}_1 b_2} \right| \leq \left| \frac{a_2 - a_1}{1 - \bar{a}_1 a_2} \right|. \tag{10.4}$$

There is a "uniqueness" result in the Schwarz–Pick lemma. If either

$$|f'(a)| = \frac{1 - |b|^2}{1 - |a|^2} \quad \text{or} \quad \left| \frac{b_2 - b_1}{1 - \bar{b}_1 b_2} \right| = \left| \frac{a_2 - a_1}{1 - \bar{a}_1 a_2} \right| \quad \text{with } a_1 \neq a_2, \tag{10.5}$$

then the function f is a conformal self-mapping (one-to-one, onto holomorphic function) of $D(0, 1)$ to itself.

We cannot discuss the verification of the Schwarz–Pick lemma right now. It depends on knowing the conformal self-maps of the disc—a topic which we shall treat later. The reader should at least observe at this time that, in (10.3), if $a = b = 0$ then the result reduces to the classical Schwarz lemma. Further, in (10.4), if $a_1 = b_1 = 0$ and $a_2 = z$, $b_2 = f(z)$, then the result reduces to the Schwarz lemma.

Exercises

1.   Let $U = \{z \in \mathbb{C} : \operatorname{Im} z > 0\}$ (the upper halfplane). Formulate and demonstrate a version of the Schwarz lemma for holomorphic functions $f : U \to U$. [**Hint:** It is useful to note that the mapping $\psi(z) = i(1 - z)/(1 + z)$ maps the unit disc to U in a holomorphic, one-to-one, and onto fashion.]

2.   Let U be as in Exercise 1. Formulate and demonstrate a version of the Schwarz lemma for holomorphic functions $f : D(0, 1) \to U$.

3.   There is no Schwarz lemma for holomorphic functions $f : \mathbb{C} \to \mathbb{C}$. Give a detailed justification for this statement. Can you suggest why the Schwarz lemma fails in this new context?

4.   Give a detailed justification for the formula

$$(f^{-1})'(w) = \frac{1}{f'(z)}.$$

Here $f(z) = w$ and f is a holomorphic function. Part of your job here is to provide suitable hypotheses about the function f.

5. Provide the details of the verification of the Schwarz–Pick lemma. [**Hint:** If $f(a) = b$, then consider $g(z) = \varphi_b \circ f \circ \varphi_{-a}$ and apply the Schwarz lemma.]

6. Let $U \subseteq \mathbb{C}$ be a domain. Let $P \in U$. Suppose that $f : U \to U$ is a holomorphic mapping with $f(P) = P$ and $f'(P) = 1$. Then it can be shown that f must be the identity mapping. This nice generalization of the Schwarz lemma, due to H. Cartan, is a bit tricky to prove. But you can at least verify it by example with U the disc, the annulus, or the unit square.

Chapter 11

The Geometric Theory of Holomorphic Functions

11.1 The Idea of a Conformal Mapping

11.1.1 Conformal Mappings

The main objects of study in this chapter are holomorphic functions $h : U \to V$, with U and V open in $\mathbb{C}$, that are one-to-one and onto. Such a holomorphic function is called a *conformal* (or *biholomorphic*) mapping. The fact that h is supposed to be one-to-one implies that h' is nowhere zero on U [remember that if h' vanishes to order $k \geq 1$ at a point $P \in U$, then h is $(k+1)$-to-1 in a small neighborhood of P—see §§9.2.1]. As a result, $h^{-1} : V \to U$ is also holomorphic—as we discussed in §§9.2.1. A conformal map $h : U \to V$ from one open set to another can be used to transfer holomorphic functions on U to V and vice versa: that is, $f : V \to \mathbb{C}$ is holomorphic if and only if $f \circ h$ is holomorphic on U; and $g : U \to \mathbb{C}$ is holomorphic if and only if $g \circ h^{-1}$ is holomorphic on V.

Thus, if there is a conformal mapping from U to V, then U and V are essentially indistinguishable from the viewpoint of complex function theory. On a practical level, one can often study holomorphic functions on a rather complicated open set by first mapping that open set to some simpler open set, then transferring the holomorphic functions as indicated.

11.1.2   Conformal Self-Maps of the Plane

The simplest open subset of $\mathbb{C}$ is $\mathbb{C}$ itself. Thus it is natural to begin our study of conformal mappings by considering the conformal mappings of $\mathbb{C}$ to itself. In fact the conformal mappings from $\mathbb{C}$ to $\mathbb{C}$ can be explicitly described as follows:

> A function $f : \mathbb{C} \to \mathbb{C}$ is a conformal mapping if and only if there are complex numbers a, b with $a \neq 0$ such that
>
> $$f(z) = az + b \ , \quad z \in \mathbb{C}. \tag{11.1}$$

One aspect of the result is fairly obvious: If $a, b \in \mathbb{C}$ and $a \neq 0$, then the map $z \mapsto az+b$ is certainly a conformal mapping of $\mathbb{C}$ to $\mathbb{C}$. In fact one checks easily that $z \mapsto (z - b)/a$ is the inverse mapping. The interesting part of the assertion is that these are in fact the only conformal maps of $\mathbb{C}$ to $\mathbb{C}$.

A generalization of this result about conformal maps of the plane is the following (consult §§5.1.3 as well as the detailed explanation in [GRK]):

> If $h : \mathbb{C} \to \mathbb{C}$ is a holomorphic function such that
>
> $$\lim_{|z| \to +\infty} |h(z)| = +\infty, \tag{11.2}$$
>
> then h is a polynomial.

In fact this last assertion is simply a restatement of the fact that if an entire function has a pole at infinity then it is a polynomial. We demonstrated that fact in §7.2. Now if $f : \mathbb{C} \to \mathbb{C}$ is conformal then it is easy to see that $\lim_{|z| \to +\infty} |f(z)| = +\infty$—for both f and f^{-1} take bounded sets to bounded sets. So f will be a polynomial. But if f has degree $k > 1$ then it will not be one-to-one: the equation $f(z) = \alpha$ will always have k roots. Thus f is a first-degree polynomial, which is what has been claimed.

Exercises

1. How many points in the plane uniquely determine a conformal self-map of the plane? That is to say, if $f : \mathbb{C} \to \mathbb{C}$ is conformal then what is the least k such that if $f(p_1) = p_1$, $f(p_2) = p_2$, ..., $f(p_k) = p_k$ then $f(z) \equiv z$?

2. Let $U = \mathbb{C} \setminus \{0\}$. What are all the conformal self-maps of U to U?

3. Let $U = \mathbb{C} \setminus \{0, 1\}$. What are all the conformal self-maps of U to U?

4. The function $f(z) = e^z$ is an onto mapping from $\mathbb{C}$ to $\mathbb{C} \setminus \{0\}$. Demonstrate this statement. The function is certainly *not* one-to-one. But it is *locally* one-to-one. Explain these assertions.

5. Refer to Exercise 4. The point i is in the image of f. Give an explicit description of the inverse of f near i.

6. The function $g(z) = z^2$ is an onto mapping from $\mathbb{C} \setminus \{0\}$ to $\mathbb{C} \setminus \{0\}$. It is certainly *not* one-to-one. But it is *locally* one-to-one. Explain these assertions.

11.2   Conformal Mappings of the Unit Disc

11.2.1   Conformal Self-Maps of the Disc

In this section we describe the set of all conformal maps of the unit disc to itself. Our first step is to determine those conformal maps of the disc to the disc that fix the origin. Let D denote the unit disc.

Let us begin by examining a conformal mapping $f : D \to D$ of the unit disc to itself and such that $f(0) = 0$. We are assuming that f is one-to-one and onto. Then, by Schwarz's lemma (§10.2), $|f'(0)| \leq 1$. This reasoning applies to f^{-1} as well, so that $|(f^{-1})'(0)| \leq 1$ or $|f'(0)| \geq 1$. We conclude that $|f'(0)| = 1$. By the uniqueness part of the Schwarz lemma, f must be a rotation. So there is a complex number ω with $|\omega| = 1$ such that

$$f(z) \equiv \omega z \quad \forall z \in D. \qquad (11.3)$$

It is often convenient to write a rotation as

$$\rho_\theta(z) \equiv e^{i\theta} z, \qquad (11.4)$$

where we have set $\omega = e^{i\theta}$ with $0 \leq \theta < 2\pi$.

We will next generalize this result to conformal self-maps of the disc that do not necessarily fix the origin.

11.2.2   Möbius Transformations

Construction of Möbius Transformations

For $a \in \mathbb{C}, |a| < 1$, we define

$$\varphi_a(z) = \frac{z - a}{1 - \overline{a}z}. \tag{11.5}$$

Then each φ_a is a conformal self-map of the unit disc.

To see this assertion, note that if $|z| = 1$, then

$$|\varphi_a(z)| = \left| \frac{z - a}{1 - \overline{a}z} \right| = \left| \frac{\overline{z}(z - a)}{1 - \overline{a}z} \right| = \left| \frac{1 - a\overline{z}}{1 - \overline{a}z} \right| = 1. \tag{11.6}$$

Thus φ_a takes the boundary of the unit disc to itself. Since $\varphi_a(0) = -a \in D$, we conclude that φ_a maps the unit disc to itself. The same reasoning applies to $(\varphi_a)^{-1} = \varphi_{-a}$, hence φ_a is a one-to-one conformal map of the disc to the disc.

The biholomorphic self-mappings of D can now be completely characterized.

11.2.3   Self-Maps of the Disc

Let $f : D \to D$ be a holomorphic function. Then f is a conformal self-map of D if and only if there are complex numbers a, ω with $|\omega| = 1, |a| < 1$ such that

$$f(z) = \omega \cdot \varphi_a(z) \quad \forall z \in D. \tag{11.7}$$

In other words, any conformal self-map of the unit disc to itself is the composition of a Möbius transformation with a rotation.

It can also be shown that any conformal self-map f of the unit disc can be written in the form

$$f(z) = \varphi_b(\eta \cdot z), \tag{11.8}$$

for some Möbius transformation φ_b and some complex number η with $|\eta| = 1$.

The reasoning is as follows: Let $f : D \to D$ be a conformal self-map of the disc and suppose that $f(0) = a \in D$. Consider the new holomorphic mapping $g = \varphi_a \circ f$. Then $g : D \to D$ is conformal and $g(0) = 0$. By what we learned in §§11.2.1, $g(z) = w \cdot z$ for some unimodular w. But this says that $f(z) = (\varphi_a)^{-1}(w \cdot z)$ or

$$f(z) = \varphi_{-a}(wz) \, .$$

That is formulation (11.8) of our result. We invite the reader to find a verification of (11.7).

EXAMPLE 8 Let us find a conformal map of the disc to the disc that takes $i/2$ to $2/3 - i/4$.

We know that $\varphi_{i/2}$ takes $i/2$ to 0. And we know that $\varphi_{-2/3+i/4}$ takes 0 to $2/3 - i/4$. Thus

$$\psi = \varphi_{-2/3+i/4} \circ \varphi_{i/2}$$

has the desired property. □

Exercises

1. Give a conformal self-map of the disc that sends $i/4 - 1/2$ to $i/3$.

2. Let a_1, a_2, b_1, b_2 be arbitrary points of the unit disc. Explain why there does not necessarily exist a holomorphic function from $D(0, 1)$ to $D(0, 1)$ such that $f(a_1) = b_1$ and $f(a_2) = b_2$.

3. Let $U = \{z \in \mathbb{C} : \operatorname{Im} z > 0\}$ (the upper halfplane). Calculate all the conformal self-mappings of U to U. [**Hint:** The function $f(z) = i(1 - z)/(1 + z)$ maps the unit disc to U conformally.]

4. Let U be as in Exercise 3. Calculate all the conformal maps of $D(0, 1)$ to U.

5. Let $P \in \mathbb{C}$ and $r > 0$. Calculate all the conformal self-maps of $D(P, r)$ to $D(P, r)$.

6. Let $U = D(0, 1) \setminus \{0\}$. Calculate all the conformal self-maps of U to U.

11.3   Linear Fractional Transformations

11.3.1   Linear Fractional Mappings

The automorphisms (that is, conformal self-mappings) of the unit disc D are special cases of functions of the form

$$z \mapsto \frac{az+b}{cz+d} \quad, \quad a,b,c,d \in \mathbb{C}. \tag{11.9}$$

It is worthwhile to consider functions of this form in generality. One restriction on this generality needs to be imposed, however; if $ad-bc = 0$, then the numerator is a constant multiple of the denominator provided that the denominator is not identically zero. So if $ad - bc = 0$, then the function is either constant or has zero denominator and is nowhere defined. Thus only the case $ad - bc \neq 0$ is worth considering in detail.

A function of the form

$$z \mapsto \frac{az+b}{cz+d} \quad, \quad ad - bc \neq 0, \tag{11.10}$$

is called a *linear fractional transformation*.

Note that $(az + b)/(cz + d)$ is not necessarily defined for all $z \in \mathbb{C}$. Specifically, if $c \neq 0$, then it is undefined at $z = -d/c$. In case $c \neq 0$,

$$\lim_{z \to -d/c} \left| \frac{az+b}{cz+d} \right| = +\infty. \tag{11.11}$$

This observation suggests that one might well, for linguistic convenience, adjoin formally a "point at ∞" to $\mathbb{C}$ and consider the value of $(az + b)/(cz + d)$ to be ∞ when $z = -d/c \, (c \neq 0)$. Thus we will think of both the domain and the range of our linear fractional transformation to be $\mathbb{C} \cup \{\infty\}$ (we sometimes also use the notation $\widehat{\mathbb{C}}$ instead of $\mathbb{C} \cup \{\infty\}$). Specifically, we are led to the following alternative method for describing a linear fractional transformation.

A function $f : \mathbb{C} \cup \{\infty\} \to \mathbb{C} \cup \{\infty\}$ is a *linear fractional transformation* if there exists $a, b, c, d \in \mathbb{C}, ad - bc \neq 0$, such that either

(a) $c = 0, d \neq 0, f(\infty) = \infty$, and $f(z) = (a/d)z + (b/d)$ for all $z \in \mathbb{C}$;

or

(b) $c \neq 0, f(\infty) = a/c, f(-d/c) = \infty$, and $f(z) = (az+b)/(cz+d)$ for all $z \in \mathbb{C}, z \neq -d/c$.

It is important to realize that, as before, the status of the point ∞ is entirely formal: we are just using it as a linguistic convenience, to keep track of the behavior of $f(z)$ both where it is not defined as a map on $\mathbb{C}$ and to keep track of its behavior when $|z| \to +\infty$. The justification for the particular devices used is the fact that

(c) $\lim_{|z| \to +\infty} f(z) = f(\infty)$ [$c = 0$; case (a) of the definition]

(d) $\lim_{z \to -d/c} |f(z)| = +\infty$ [$c \neq 0$; case (b) of the definition].

11.3.2   The Topology of the Extended Plane

The limit properties of f that we described in §§11.3.1 can be considered as continuity properties of f from $\mathbb{C} \cup \{\infty\}$ to $\mathbb{C} \cup \{\infty\}$ using the definition of continuity that comes from the topology on $\mathbb{C} \cup \{\infty\}$ (which we are about to define). It is easy to formulate that topology in terms of open sets. But it is also convenient to formulate that same topological structure in terms of convergence of sequences:

A sequence $\{p_i\}$ in $\mathbb{C} \cup \{\infty\}$ *converges to* $p_0 \in \mathbb{C} \cup \{\infty\}$ (notation $\lim_{i \to \infty} p_i = p_0$) if either

(e) $p_0 = \infty$ and $\lim_{i \to +\infty} |p_i| = +\infty$ where the limit is taken for all i such that $p_i \in \mathbb{C}$;

or

(f) $p_0 \in \mathbb{C}$, all but a finite number of the p_i are in $\mathbb{C}$ and $\lim_{i \to \infty} p_i = p_0$ in the usual sense of convergence in $\mathbb{C}$.

11.3.3   The Riemann Sphere

Stereographic projection puts $\widehat{\mathbb{C}} = \mathbb{C} \cup \{\infty\}$ into one-to-one correspondence with the two-dimensional sphere S in $\mathbb{R}^3, S = \{(x, y, z) \in \mathbb{R}^3 : x^2 + y^2 + z^2 = 1\}$ in such a way that the topology is preserved in both directions of the correspondence.

In detail, begin by imagining the unit sphere bisected by the complex plane with the center of the sphere $(0, 0, 0)$ coinciding with the origin in

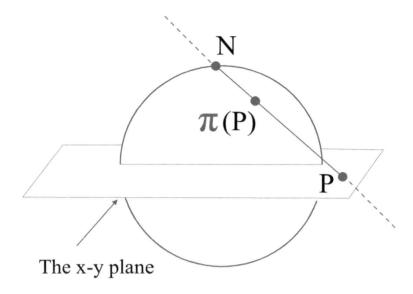

The x-y plane

Figure 11.1: Stereographic projection.

the plane—see Figure 11.1. We define the stereographic projection as follows: If $P = (x, y) \in \mathbb{C}$, then connect P to the "north pole" N of the sphere with a line segment. The point $\pi(P)$ of intersection of this segment with the sphere is called the *stereographic projection* of P. Note that, under stereographic projection, the "point at infinity" in the plane corresponds to the north pole N of the sphere. For this reason, $\mathbb{C} \cup \{\infty\}$ is often thought of as "being" a sphere, and is then called, for historical reasons, the *Riemann sphere*.

The construction we have just described is another way to think about the "extended complex plane"—see §§11.3.2. In these terms, linear fractional transformations become homeomorphisms of $\mathbb{C} \cup \{\infty\}$ to itself. (Recall that a *homeomorphism* is, by definition, a one-to-one, onto, continuous mapping with a continuous inverse.)

If $f : \mathbb{C} \cup \{\infty\} \to \mathbb{C} \cup \{\infty\}$ is a linear fractional transformation, then f is a one-to-one, onto, continuous function. Also, $f^{-1} : \mathbb{C} \cup \{\infty\} \to \mathbb{C} \cup \{\infty\}$ is a linear fractional transformation, and is thus a one-to-one, onto, continuous function.

If $g : \mathbb{C} \cup \{\infty\} \to \mathbb{C} \cup \{\infty\}$ is also a linear fractional transformation, then $f \circ g$ is a linear fractional transformation.

The simplicity of language obtained by adjoining ∞ to $\mathbb{C}$ (so that the composition and inverse properties of linear fractional transformations obviously hold) is well worth the trouble. Certainly one does not wish to consider the multiplicity of special possibilities when composing $(Az + B)/(Cz + D)$ with $(az + b)/(cz + d)$ (namely $c = 0, c \neq 0, aC + cD \neq 0, aC + cD = 0$, etc.) that arise every time composition is considered.

In fact, it is worth summarizing what we have learned in a theorem (see §§11.3.4). First note that it makes sense now to talk about a homeomorphism from $\mathbb{C} \cup \{\infty\}$ to $\mathbb{C} \cup \{\infty\}$ being conformal: this just means that it (and hence its inverse) are holomorphic in our extended sense. More precisely, a function g is holomorphic at the point ∞ if $g(1/z)$ is holomorphic at the origin. A function h which takes the value ∞ at p is holomorphic at p if $1/h$ is holomorphic at p.

If φ is a conformal map of $\mathbb{C} \cup \{\infty\}$ to itself, then, after composing with a linear fractional transformation, we may suppose that φ maps ∞ to itself. Thus φ, after composition with a linear fraction transformation, is linear. It follows that φ itself is linear fractional. The following result summarizes the situation:

11.3.4   Conformal Self-Maps of the Riemann Sphere

THEOREM 4 *A function φ is a conformal self-mapping of $\widehat{\mathbb{C}} = \mathbb{C} \cup \{\infty\}$ to itself if and only if φ is linear fractional.*

We turn now to the actual utility of linear fractional transformations (beyond their having been the form of automorphisms of D—see §§11.2.1–11.2.3—and the form of all conformal self maps of $\mathbb{C} \cup \{\infty\}$ to itself in the present section). One of the most frequently occurring uses is the following:

11.3.5   The Cayley Transform

The Cayley Transform The linear fractional transformation $z \mapsto (i - z)/(i + z)$ maps the upper halfplane $\{z : \mathrm{Im}\, z > 0\}$ conformally onto the unit disc $D = \{z : |z| < 1\}$.

11.3.6   Generalized Circles and Lines

Calculations of the type that we have been discussing are straightforward but tedious. It is thus worthwhile to seek a simpler way to understand what the image under a linear fractional transformation of a given region is. For regions bounded by line segments and arcs of circles the following result gives a method for addressing this issue:

Let $\mathcal{C}$ be the set of subsets of $\mathbb{C} \cup \{\infty\}$ consisting of (i) circles and (ii) sets of the form $L \cup \{\infty\}$ where L is a line in $\mathbb{C}$. We call the elements of $\mathcal{C}$ "generalized circles." Then every linear fractional transformation φ takes elements of $\mathcal{C}$ to elements of $\mathcal{C}$. One verifies this last assertion by noting that any linear fractional transformation is the composition of dilations, translations, and the inversion map $z \mapsto 1/z$; and each of these component maps clearly sends generalized circles to generalized circles.

11.3.7   The Cayley Transform Revisited

To illustrate the utility of this last result, we return to the Cayley transformation

$$z \mapsto \frac{i - z}{i + z}. \tag{11.12}$$

Under this mapping the point ∞ is sent to -1, the point 1 is sent to $(i-1)/(i+1) = i$, and the point -1 is sent to $(i-(-1))/(i+(-1)) = -i$. Thus the image under the Cayley transform (a linear fractional transformation) of three points on $\mathbb{R} \cup \{\infty\}$ contains three points on the unit circle. Since three points determine a (generalized) circle, and since linear fractional transformations send generalized circles to generalized circles, we may conclude that the Cayley transform sends the real line to the unit circle. Now the Cayley transform is one-to-one and onto from $\mathbb{C} \cup \{\infty\}$ to $\mathbb{C} \cup \{\infty\}$. By continuity, it either sends the upper halfplane to the (open) unit disc or to the complement of the closed unit disc. The image of i is 0, so in fact the Cayley transform sends the upper halfplane to the unit disc.

11.3.8 Summary Chart of Linear Fractional Transformations

The next chart summarizes the properties of some important linear fractional transformations. Note that $U = \{z \in \mathbb{C} : \operatorname{Im} z > 0\}$ is the upper halfplane and $D = \{z \in \mathbb{C} : |z| < 1\}$ is the unit disc; the domain variable is z and the range variable is w.

Linear Fractional Transformations

Domain	Image	Conditions	Formula
$z \in \widehat{\mathbb{C}}$	$w \in \widehat{\mathbb{C}}$		$w = \frac{az+b}{cz+d}$
$z \in D$	$w \in U$		$w = i \cdot \frac{1-z}{1+z}$
$z \in U$	$w \in D$		$w = \frac{i-z}{i+z}$
$z \in D$	$w \in D$		$w = \frac{z-a}{1-\bar{a}z}$, $\quad |a| < 1$
$z \in \mathbb{C}$	$w \in \mathbb{C}$	$L(z_1) = w_1$ $L(z_2) = w_2$ $L(z_3) = w_3$	$L(z) = S^{-1} \circ T$ $T(z) = \frac{z-z_1}{z-z_3} \cdot \frac{z_2-z_3}{z_2-z_1}$ $S(m) = \frac{m-w_1}{m-w_3} \cdot \frac{w_2-w_3}{w_2-w_1}$

Exercises

1. Calculate the inverse of the Cayley transform.

2. Calculate all the conformal mappings of the unit disc to the upper halfplane.

3. Calculate all the conformal mappings from $U = \{z \in \mathbb{C} : \operatorname{Re}((3 - i) \cdot z) > 0\}$ to $V = \{z \in \mathbb{C} : \operatorname{Re}((4 + 2i) \cdot z > 0\}$.

4. Calculate all the conformal mappings from the disc $D(p, r)$ to the disc $D(P, R)$.

5. How many points in the Riemann sphere uniquely determine a linear fractional transformation?

6. Show that a linear fractional transformation

$$\varphi(z) = \frac{az + b}{cz + d}$$

preserves the upper halfplane if and only if $ad - bc > 0$.

7. Which linear fractional transformations preserve the real line? Which preserve the unit circle?

8. Let ℓ be a linear fractional transformation and C a circle in the plane. What is a quick test to determine whether ℓ maps C to another circle (rather than a line)?

9. Let ℓ be a linear fractional transformation and L a line in the plane. What is a quick test to determine whether ℓ maps L to another line (rather than a circle)?

11.4   The Riemann Mapping Theorem

11.4.1   The Concept of Homeomorphism

Two open sets U and V in $\mathbb{C}$ are *homeomorphic* if there is a one-to-one, onto, continuous function $f : U \to V$ with $f^{-1} : V \to U$ also continuous. Such a function f is called a *homeomorphism* from U to V (see also §§11.3.3).

11.4.2   The Riemann Mapping Theorem

The Riemann mapping theorem, sometimes called the greatest theorem of the nineteenth century, asserts in effect that any planar domain (other than $\mathbb{C}$ itself) that has the topology of the unit disc also has the conformal structure of the unit disc. Even though this theorem has been subsumed by the great uniformization theorem of Köbe (see [FAK]), it is still striking in its elegance and simplicity:

If U is an open subset of $\mathbb{C}, U \neq \mathbb{C}$, and if U is homeomorphic to D, then U is conformally equivalent to D. That is, there is a holomorphic mapping $\psi : U \to D$ which is one-to-one and onto.

11.4.3 The Riemann Mapping Theorem: Second Formulation

An alternative formulation of this theorem uses the concept of "simply connected" (see also §§4.2.2). We say that a connected open set U in the complex plane is simply connected if any closed curve in U can be continuously deformed to a point. (This is just a precise way of saying that U has no holes. Yet another formulation of the notion is that the complement of U has only one connected component—see [GRK].)

> **Theorem:** If U is an open subset of $\mathbb{C}, U \neq \mathbb{C}$, and if U is simply connected, then U is conformally equivalent to D.

Exercises

1. The Riemann mapping theorem is an astonishing result. One corollary is that any simply connected open set in the plane is homeomorphic to the plane. Explain, remembering that the Riemann mapping theorem does not apply to the case when the domain in question is the entire plane.

2. Explain why the Riemann mapping theorem must exclude the entire plane as a candidate for being conformally equivalent to the unit disc.

3. Let $U \subseteq \mathbb{C}$ be a domain and a proper subset that is simply connected. Let $a \in U$. Show that there is a unique conformal mapping φ of the unit disc $D(0, 1)$ to U with the property that $\varphi(0) = a$ and $\varphi'(0) > 0$.

4. Let $U \subseteq \mathbb{C}$ be a proper subset that is simply connected. Let $a, b \in U$ be arbitrary elements. Explain why there is not necessarily a conformal mapping $\varphi : D(0, 1) \to U$ such that $\varphi(0) = a$ and $\varphi(1/2) = b$. Give an explicit example where there is no mapping.

5. Let $A = \{z \in \mathbb{C} : 1/2 < |z| < 2\}$. Define a holomorphic function φ on A by $\varphi(z) = z + 1/z$. Explain why this is a mapping of A onto the

interior of an ellipse. What ellipse is it? Why does this example not contradict the dictum that linear fractional transformations take lines and circles to lines and circles?

6. The Riemann mapping theorem guarantees (abstractly) that there is a conformal map of the strip $\{z \in \mathbb{C} : |\text{Im } z| < 1\}$ onto the unit disc. Write down this mapping explicitly.

7. The Riemann mapping theorem guarantees (abstractly) that there is a conformal map of the quarter-plane $\{z = x + iy \in \mathbb{C} : x > 0, y > 0\}$. Write down this mapping explicitly.

8. Calculate all the conformal self-mappings of the strip $\{z \in \mathbb{C} : |\text{Im } z| < 1\}$.

11.5   Conformal Mappings of Annuli

11.5.1   A Riemann Mapping Theorem for Annuli

The Riemann mapping theorem tells us that, from the point of view of complex analysis, there are only two simply connected planar domains: the disc and the plane. Any other simply connected region is biholomorphic to one of these. It is natural then to ask about domains with holes. Take, for example, a domain U with precisely one hole. Is it conformally equivalent to an annulus?

   Note that, if $c > 0$ is a constant, then for any $R_1 < R_2$ the annuli

$$A_1 \equiv \{z : R_1 < |z| < R_2\} \quad \text{and} \quad A_2 \equiv \{z : cR_1 < |z| < cR_2\} \quad (11.13)$$

are biholomorphically equivalent under the mapping $z \mapsto cz$. The surprising fact that we shall learn is that these are the *only* circumstances under which two annuli are equivalent:

11.5.2   Conformal Equivalence of Annuli

Let

$$A_1 = \{z \in \mathbb{C} : 1 < |z| < R_1\} \quad (11.14)$$

and

$$A_2 = \{z \in \mathbb{C} : 1 < |z| < R_2\}. \quad (11.15)$$

Then A_1 is conformally equivalent to A_2 if and only if $R_1 = R_2$.

A perhaps more striking result, and more difficult to demonstrate, is this:

> Let $U \subseteq \mathbb{C}$ be any bounded domain with *one hole*—this means that the complement of U has two connected components, one bounded and one not. Then U is conformally equivalent to some annulus.

The verifications of these results are rather deep and difficult. We cannot discuss them in any detail here, but include their statements for completeness. See [AHL], [GRK], and [KRA4] for discursive discussions of these theorems.

11.5.3 Classification of Planar Domains

The classification of planar domains up to biholomorphic equivalence is a part of the theory of Riemann surfaces. For now, we comment that one of the startling classification theorems (a generalization of the Riemann mapping theorem) is that any bounded planar domain with finitely many "holes" is conformally equivalent to the unit disc with finitely many closed circular arcs, coming from circles centered at the origin, removed. See Figure 11.2. (Here a "hole" in the present context means a bounded, connected component of the complement of the domain in $\mathbb{C}$, a concept which coincides with the intuitive idea of a hole.) An alternative equivalent statement is that any bounded planar domain with finitely many holes is conformally equivalent to the plane with finitely many vertical slits centered on the x-axis removed (see [AHL] or [KRA4]). Refer to Figure 11.3. The analogous result for domains with infinitely many holes is known to be true when the number of holes is countable (see [HES]).

Exercises

1. How much data is needed to uniquely determine a conformal mapping of annuli? Suppose that $A_1 = \{z \in \mathbb{C} : 1/2 < |z| < 2\}$ and $A_2 = \{z \in \mathbb{C} : 1 < |z| < 4\}$. Say that f is a conformal mapping of A_1 to A_2 such that $f(1) = 2$. Is there only one such mapping?

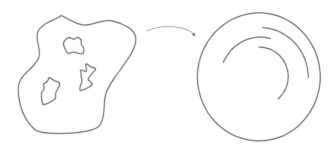

Figure 11.2: Representation of a domain on the disc with circular arcs removed.

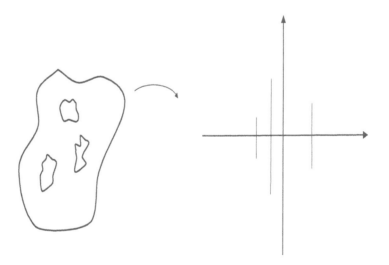

Figure 11.3: Representation of a domain on the plane with vertical slits removed.

2. Define the annulus $A = \{z \in \mathbb{C} : 1/2 < |z| < 2\}$. Certainly any rotation is a conformal self-mapping of A. Also the inversion $\psi : z \mapsto 1/z$ is a conformal mapping of A to itself. Verify these assertions. Can you think of any other conformal mappings of A to A?

3. Let A be an annulus and ℓ a linear fractional transformation. What can the image of A under ℓ be? Describe all the possibilities.

4. What is the image of the halfplane $\{z \in \mathbb{C} : \operatorname{Re} z < 0\}$ under the mapping $z \mapsto e^z$? Is the mapping one-to-one?

5. What is the image of the strip $\{z \in \mathbb{C} : 1 < \operatorname{Re} z < 2\}$ under the mapping $z \mapsto e^z$? Is the mapping one-to-one?

6. Consider the region $U = \widehat{\mathbb{C}} \setminus \{z \in \mathbb{C} : \operatorname{Im} z = 0, 0 < \operatorname{Re} z < 1\}$. The image of U under the mapping $z \mapsto 1/z$ is the slit plane $V = \mathbb{C} \setminus \{z \in \mathbb{C} : \operatorname{Im} z = 0, 0 < \operatorname{Re} z < 1\}$. Verify this assertion. Draw a picture of the domain and range of this function. Now apply the mapping $z \mapsto \sqrt{z}$ to V. The result is a halfplane. Finally, a suitable Cayley transform will take that last halfplane to the unit disc. Thus the original region $U \subseteq \widehat{\mathbb{C}}$ is conformally equivalent to the unit disc.

7. It is a fact that a conformal self-mapping of *any* planar domain that has three fixed points is the identity mapping (see [FIF], [LES], [MAS]). Show that there is a non-trivial conformal self-map of the annulus $A = \{z \in \mathbb{C} : 1/2 < |z| < 2\}$ having two distinct fixed points.

8. Refer to the last exercise for background. Show that any conformal self-map of the disc having two distinct fixed points is in fact the identity. How many distinct fixed boundary points of the disc will force the mapping to be the identity?

11.6 A Compendium of Useful Conformal Mappings

Here we present a graphical compendium of commonly used conformal mappings. Most of the mappings that we present here are given by explicit formulas, and are also represented in figures. Wherever possible, we also provide the inverse of the mapping.

In each case, the domain of the mapping is in the z-plane, with $x + iy = z \in \mathbb{C}$. And the range of the mapping is in the w-plane, with $u + iv = w \in \mathbb{C}$. In some examples, it is appropriate to label special points $a, b, c, \ldots$ in the domain of the mapping and then to specify their corresponding image points $A, B, C, \ldots$ in the range. In other words, if the mapping is called f, then $f(a) = A$, $f(b) = B$, $f(c) = C$, etc.

All of the mappings presented here map the shaded region in the z-plane *onto* the shaded region in the w-plane. Most of the mappings are one-to-one (i.e., they do *not* map two distinct points in the domain of the mapping to the same point in the range of the mapping). In a few cases the mapping is *not* one-to-one; these few examples will be clear from context.

In the majority of these examples, the mapping is given by an explicit formula. In a few cases, such as the Schwarz–Christoffel mapping, the mapping is given by a semi-explicit integral. In one example (Figure 11.4), the mapping is given by an elliptic integral. Such integrals cannot be evaluated in closed form. But they can be calculated to any degree of accuracy using methods of numerical integration. The book [KOB] gives an extensive listing of explicit conformal mappings; see also [CCP]. The book [NEH] is a classic treatise on the theory of conformal mappings.

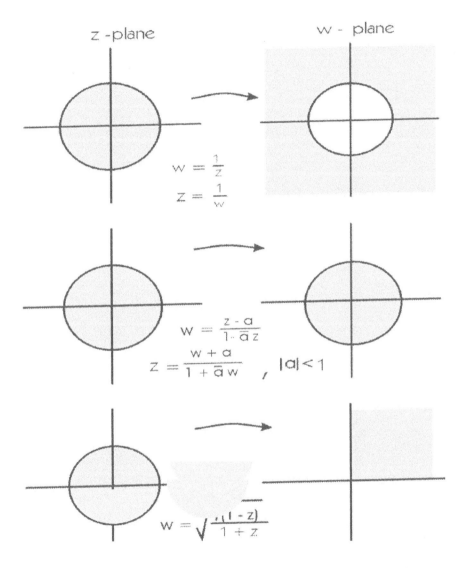

Figure 11.4: **(top)** Map of the disc to its complement; **(middle)** map of the disc to the disc; **(bottom)** map of the upper half-disc to the first quadrant.

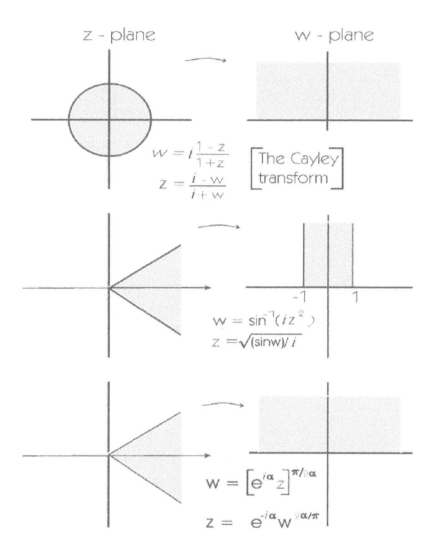

Figure 11.5: **(top)** The Cayley transform: A map of the disc to the upper halfplane; **(middle)** map of a cone to a halfstrip; **(bottom)** map of a cone to the upper halfplane.

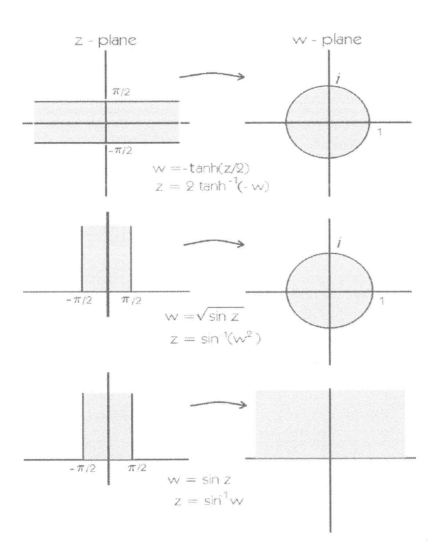

Figure 11.6: **(top)** Map of a strip to the disc; **(middle)** map of a halfstrip to the disc; **(bottom)** map of a halfstrip to the upper halfplane.

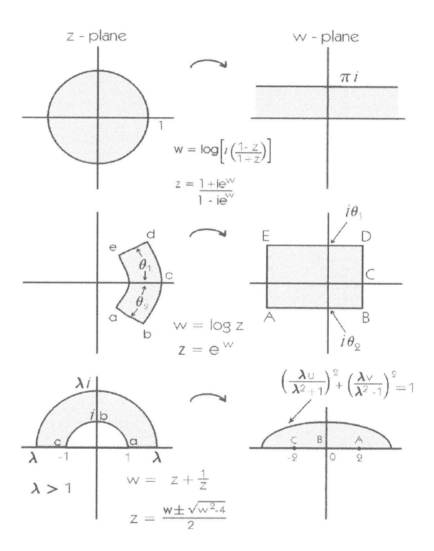

Figure 11.7: **(top)** Map of the disc to a strip; **(middle)** map of an annular sector to the interior of a rectangle; **(bottom)** map of a half-annulus to the interior of a half-ellipse.

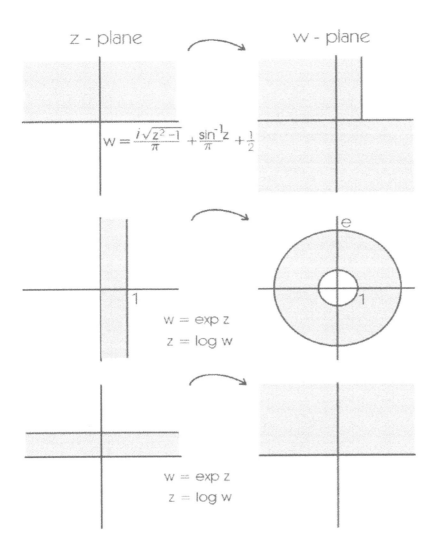

Figure 11.8: **(top)** Map of the upper halfplane to a 3/4-plane; **(middle)** map of a strip to an annulus; **(bottom)** map of a strip to the upper halfplane.

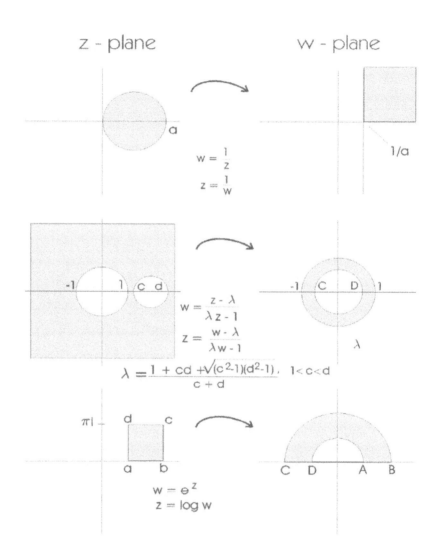

Figure 11.9: **(top)** Map of a disc to a quadrant; **(middle)** map of the complement of two discs to an annulus; **(bottom)** map of the interior of a rectangle to a half-annulus.

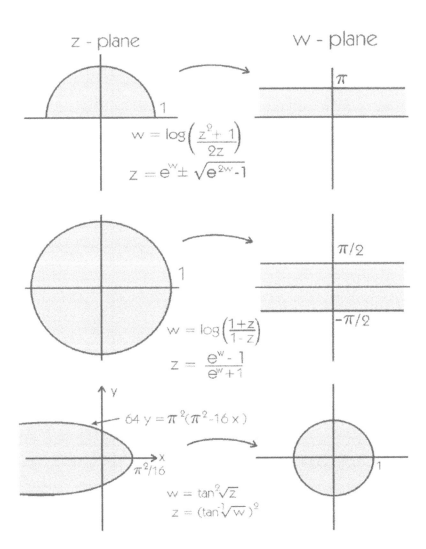

Figure 11.10: **(top)** Map of a half-disc to a strip; **(middle)** map of a disc to a strip; **(bottom)** map of the inside of a parabola to a disc.

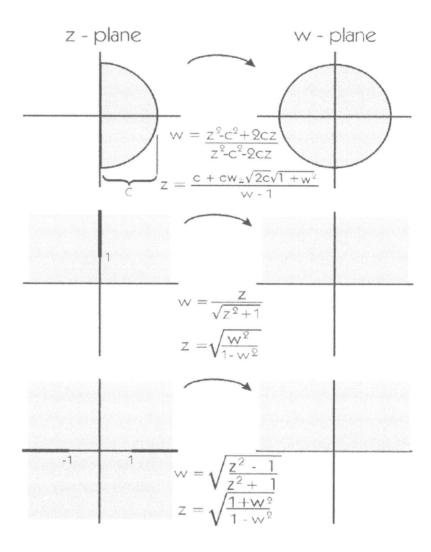

Figure 11.11: **(top)** Map of a half-disc to a disc; **(middle)** map of the slotted upper halfplane to upper halfplane; **(bottom)** map of the double-sliced plane to the upper halfplane.

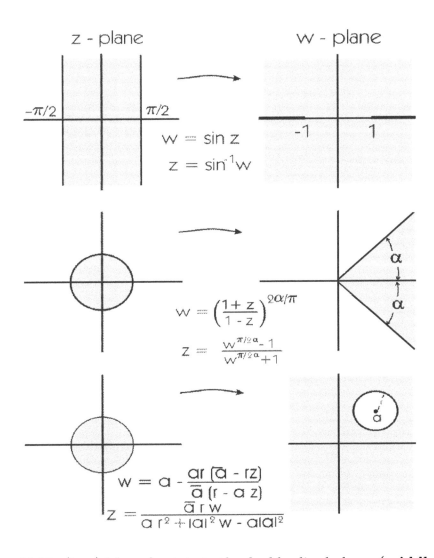

Figure 11.12: **(top)** Map of a strip to the double-sliced plane; **(middle)** map of the disc to a cone; **(bottom)** map of a disc to the complement of a disc.

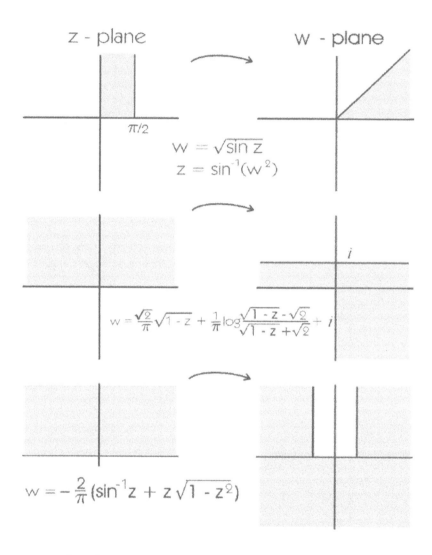

Figure 11.13: **(top)** Map of a halfstrip to a half-quadrant; **(middle)** map of the upper halfplane to a right-angle region; **(bottom)** map of the upper halfplane to the plane less a halfstrip.

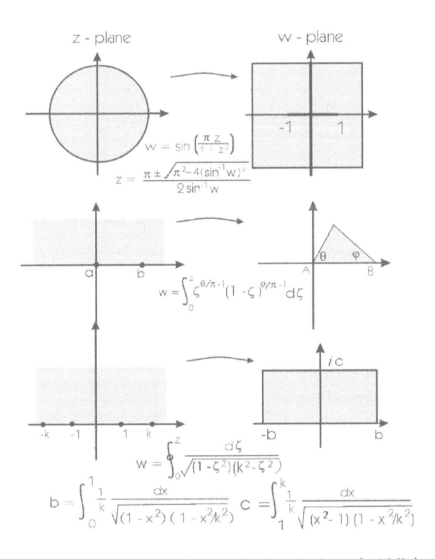

Figure 11.14: **(top)** Map of the disc to the slotted plane; **(middle)** map of the upper halfplane to the interior of a triangle; **(bottom)** map of the upper halfplane to the interior of a rectangle.

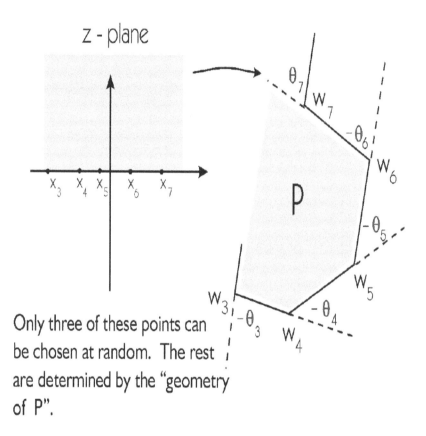

Only three of these points can
be chosen at random. The rest
are determined by the "geometry
of P".

$$w = f(z) = A \int_0^z (\zeta - x_1)^{\theta_1/\pi} \cdot (\zeta - x_2)^{\theta_2/\pi} \cdots (\zeta - x_3)^{\theta_{n-1}/\pi} d\zeta + B$$

$$f(x_1) = w_1, f(x_2) = w_2, \ldots, f(x_n) = w_n$$

Figure 11.15: The Schwarz–Christoffel formula.

Chapter 12

Applications that Depend on Conformal Mapping

12.1 Conformal Mapping

12.1.1 The Study of Conformal Mappings

Part of the utility of conformal mappings is that they can be used to transform a problem on a given domain V to another domain U (see also §§11.1.1). Often we take U to be a standard domain such as the disc

$$D = \{z \in \mathbb{C} : |z| < 1\} \qquad (12.1)$$

or the upper halfplane

$$U = \{z \in \mathbb{C} : \operatorname{Im} z > 0\}. \qquad (12.2)$$

Particularly in the study of partial differential equations, it is important to have an *explicit* conformal mapping between the two domains. At the end of Chapter 11 we gave a useful and explicit compendium of conformal mappings.

The reader will find that, even in cases where the precise mapping that he/she seeks has not been listed, he can (much as with a table of integrals) combine several of the given mappings to produce the results that he/she seeks. It is also the case that the techniques presented can be modified to suit a variety of different situations.

The references [KOB] and [CCP] give more comprehensive lists of conformal mappings.

12.2   Application of Conformal Mapping to the Dirichlet Problem

12.2.1   The Dirichlet Problem

Let $\Omega \subseteq \mathbb{C}$ be a domain whose boundary consists of finitely many smooth curves. The *Dirichlet problem* (see §12.2, §13.3, §16.1, which is a mathematical problem of interest in its own right, is the boundary value problem

$$\begin{aligned} \triangle u &= 0 \quad &&\text{on } \Omega \\ u &= f \quad &&\text{on } \partial\Omega. \end{aligned} \tag{12.3}$$

The way to think about this problem is as follows: a continuous data function f on the boundary of the domain is given. To solve the corresponding Dirichlet problem, one seeks a continuous function u on the closure of Ω (i.e., the union of Ω and its boundary) such that u is harmonic on the interior Ω and u agrees with f on the boundary. We shall now describe three distinct physical situations that are mathematically modeled by the Dirichlet problem.

12.2.2   Physical Motivation for the Dirichlet Problem

I. Heat Diffusion: Imagine that Ω is a thin plate of heat-conducting metal. One applies some initial temperature to the boundary points. We let the function f describe that initial heat distribution on the boundary.

The shape of Ω is arbitrary (not necessarily a rectangle). See Figure 12.1. A function $u(x, y)$ describes the temperature at each point (x, y) in Ω. It is a standard situation in engineering or physics to consider idealized heat sources or sinks that maintain specified (fixed) values of u on certain parts of the boundary; other parts of the boundary are to be thermally insulated. One wants to find the steady-state heat distribution on Ω that is determined by the given boundary conditions. If we let f denote the temperature specified on the boundary, then it turns out the solution of the Dirichlet problem (12.3) is the function that describes the heat distribution (see [COH], [KRA4], [BCH, p.300] and references therein for a derivation of this mathematical model for heat distribution).

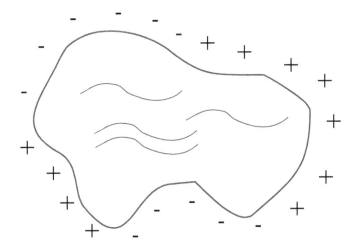

Figure 12.1: Heat distribution on the edge of a metal plate.

We will present below some specific examples of heat diffusion problems that illustrate the mathematical model that we have discussed here, and we will show how conformal mapping can be used in aid of the solutions of the problems.

II. Electrostatic Potential: Now we describe a situation in electrostatics that is modeled by the boundary value problem (12.3).

Imagine a long, hollow cylinder made of a thin sheet of some conducting material, such as copper. Split the cylinder lengthwise into two equal pieces (Figure 12.2). Separate the two pieces with strips of insulating material (Figure 12.3). Now ground the upper of the two semi-cylindrical pieces to potential zero, and keep the lower piece at some non-zero fixed potential. For simplicity in the present discussion, let us say that this last fixed potential is 1.

Note that, in the figures, the axis of the cylinder is the z-axis. Consider a slice of this cylindrical picture which is taken by setting z equal to a small constant (we want to stay away from the ends of the cylinder, where the analysis will be a bit different).

Once we have fixed a value of z, then we may study the electrostatic potential $V(x, y)$, $x^2 + y^2 < 1$, at a point inside the cylinder. Observe that $V = 0$ on the "upper" half of the circle ($y > 0$) and $V = 1$ on the "lower" half of the circle ($y < 0$)—see Figure 12.4. Physical analysis (see [BCH, p. 310]) shows that this is another Dirichlet problem, as in (12.3).

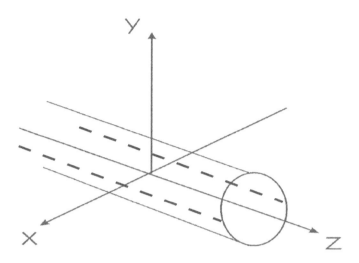

Figure 12.2: Electrostatic potential illustrated with a split cylinder.

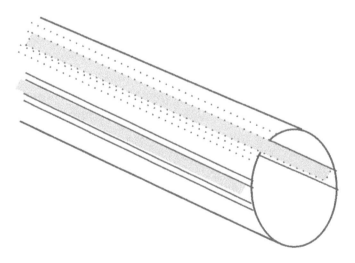

Figure 12.3: The cylindrical halves separated with insulating material.

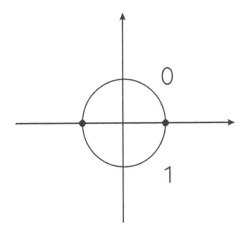

Figure 12.4: Distribution of the electrical potential.

We wish to find a harmonic function V on the disc $\{(x, y) : x^2 + y^2 \leq 1\}$ which agrees with the given potentials on the boundary.

Conformal mapping can be used as an aid in solving the problem posed here, and we shall discuss its solution below.

III. Incompressible Fluid Flow: For the mathematical model considered here, we consider a two-dimensional flow of a fluid that is

- incompressible

- irrotational

- free from viscosity

The first of these stipulations means that the fluid is of constant density, the second means that the curl is zero, and the third means that the fluid flows freely.

Identifying a point (a, b) in the x-y plane with the complex number $a + ib$ as usual, we let

$$V(x, y) = p(x, y) + iq(x, y)$$

represent the velocity vector of our fluid flow at a point (x, y). We assume that the fluid flow has no sources or sinks; and we hypothesize that p and q are C^1, or once continuously differentiable (see §§3.1.1).

The *circulation* of the fluid along any curve γ is the line integral

$$\int_\gamma V_T(x, y)\, d\sigma.$$

Here V_T represents the tangential component of the velocity along the curve γ and σ denotes arc length. We know from advanced calculus that the circulation can be written

$$\int_\gamma p(x, y)\, dx + \int_\gamma q(x, y)\, dy.$$

We assume here that γ is a positively (counterclockwise) oriented simple, closed curve that lies in a simply connected region D of the flow.

Now Green's theorem allows us to rewrite this last expression for the circulation as

$$\iint_R [q_x(x, y) - p_y(x, y)]\, dA.$$

Here the subscripts x and y represent partial derivatives, R is the region surrounded by γ, and dA is the element of area. In summary

$$\int_\gamma V_T(x, y)\, d\sigma = \iint_R [q_x(x, y) - p_y(x, y)]\, dA.$$

Let us specialize to the case that γ is a circle of radius r with center $p_0 = (x_0, y_0)$. Then the mean angular speed of the flow along γ is

$$\frac{1}{\pi r^2} \iint_R \frac{1}{2}[q_x(x, y) - p_y(x, y)]\, dA.$$

This expression also happens to represent the average of the function

$$\omega(x, y) = \frac{1}{2}[q_x(x, y) - p_y(x, y)] \tag{12.4}$$

over R. Since ω is continuous, the limit as $r \to 0$ of (12.4) is just $\omega(p_0)$. It is appropriate to call ω the *rotation* of the fluid, since it is the limit at the point p_0 of the angular speed of a circular element of the fluid at the point p_0. Since our fluid is irrotational, we set $\omega = 0$. Thus we know that

$$p_y = q_x$$

in the region D where the flow takes place. Multi-dimensional calculus then tells us that the flow is path-independent: If $X = (x, y)$ is any point in the region and γ is any path joining p_0 to X, then the integral

$$\int_\gamma p(s, t)\, ds + \int_\gamma q(s, t)\, dt$$

is independent of the choice of γ. As a result, the function

$$\varphi(x, y) = \int_{p_0}^{X} p(s, t)\, ds + q(s, t)\, dt$$

is well-defined on D. Differentiating the equation that defines φ, we find that

$$\frac{\partial}{\partial x}\varphi(x, y) = p(x, y) \quad \text{and} \quad \frac{\partial}{\partial y}\varphi(x, y) = q(x, y). \tag{12.5}$$

We call φ a *potential function* for the flow. To summarize, we know that $\nabla\varphi = (p, q)$.

The natural physical requirement that the incompressible fluid enter or leave an element of volume only by flowing through the boundary of that element (no sources or sinks) entails the mathematical condition that φ be harmonic. Thus

$$\varphi_{xx} + \varphi_{yy} = 0 \tag{12.6}$$

on D. In conclusion, studying a fluid flow with specified boundary data will entail solving the boundary value problem (12.3).

Exercises

1. Suppose that a unit disc of aluminum has initial temperature 10 degrees on the upper half of the boundary and initial temperature -10 degrees on the lower half of the boundary. Let the heat propagate throughout the disc. What is the eventual temperature at the origin? Why?

2. Exercise 1 illustrates a general principle for solutions of the Dirichlet problem, and for harmonic functions in general. Namely, that the average of the solution function u over a disc $D(P, r)$ equals the value of u at the center of the disc. Explain why this should be true in terms of the second law of thermodynamics.

3. If u is the real-valued solution of a Dirichlet problem on the unit disc then there will exist another real-valued function v on the disc so that $h = u + iv$ is holomorphic on the disc. Explain why this is so.

4. Explain why the solution of a Dirichlet problem is unique.

5. Consider a Dirichlet problem with initial boundary data that is real-valued. Then the solution u of that Dirichlet problem will also be real-valued. Explain why.

6. Refer to Exercises 1 and 5. Give an example of Dirichlet data on the unit disc that is complex-valued, but so that the solution of the Dirichlet problem at the origin is real-valued.

7. Show how to use the Cayley transform to transfer a Dirichlet problem from the unit disc to the upper half plane. In particular, what does the Dirichlet problem in Exercise 1 become on the upper half plane?

8. Refer to Exercise 7. Use the ideas there to consider the Dirichlet problem on the first quadrant. Suppose that you have Dirichlet boundary data on the first quadrant consisting of temperature 10 degrees on the positive real axis and temperature -10 degrees on the positive imaginary axis. What can you say about the solution u of this Dirichlet problem? Will u take the value 0 anywhere?

9. Consider a Dirichlet problem on the unit disc with boundary data $\varphi(e^{i\theta}) = e^{ij\theta}$, $0 \le \theta < 2\pi$ and $j = 0, 1, 2, \ldots$. Give an explicit formula for the solution of this Dirichlet problem.

10. Consider a Dirichlet problem on the unit disc with boundary data $\varphi(e^{i\theta}) = e^{ij\theta}$, $0 \le \theta < 2\pi$ and $j = -1, -2, -3, \ldots$. Give an explicit formula for the solution of this Dirichlet problem.

12.3   Physical Examples Solved by Means of Conformal Mapping

In this section we give a concrete illustration of the solution of each of the physical problems described in the last section.

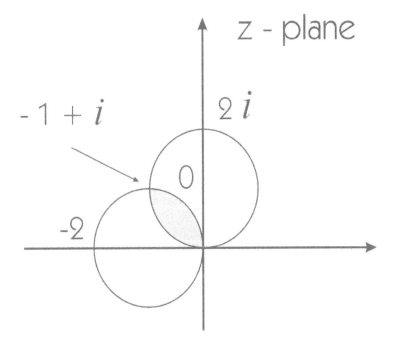

Figure 12.5: A lens-shaped piece of heat-conducting metal.

12.3.1 Steady-State Heat Distribution on a Lens-Shaped Region

EXAMPLE 9 Imagine a lens-shaped piece of heat-conducting metal as in Figure 12.5. Suppose that the initial distribution of heat is specified to be 1 on the lower boundary of the lens and 0 on the upper boundary of the lens (as illustrated in the figure). Determine the steady-state heat distribution. □

Solution: Our strategy is to use a conformal mapping to transfer the problem to a new domain on which it is easier to work. We let $z = x + iy$ denote the variable in the lens-shaped region and $w = u + iv$ denote the variable in the new region (which will be an angular region).

In fact let us construct the conformal mapping with our bare hands. If we arrange for the mapping to be linear fractional and to send the origin to the origin and the point $-1 + i$ to infinity, then (since linear fractional transformations send lines and circles to lines and circles), the images of the two circular arcs will both be lines. Let us in fact examine the mapping

$$w = f(z) = \frac{-z}{z - (-1 + i)}.$$

(The minus sign in the numerator is introduced for convenience.)

We see that $f(0) = 0$, $f(2i) = -1 - i$, and $f(-2) = -1 + i$. Using conformality (preservation of right angles), we conclude that the image of the lens-shaped region in the z-plane is the angular region in the w-plane depicted in Figure 12.6. The figure also shows on which part of the boundary the function we seek is to have value 0 and on which part it is to have value 1. It is easy to write down a harmonic function of the w variable that satisfies the required boundary conditions:

$$\varphi(w) = \frac{2}{\pi} \left(\arg w + \frac{\pi}{4} \right)$$

will certainly do the job if we demand that $-\pi < \arg w < \pi$. But then the function

$$u(z) = \varphi \circ f(z)$$

is a harmonic function on the lens-shaped domain in the z-plane that has the requisite boundary values (we use of course the fact that the composition of a harmonic function with a holomorphic function is still harmonic).

In other words, the solution to the problem originally posed on the lens-shaped domain in the z-plane is

$$u(z) = \frac{2}{\pi} \cdot \arg \left(\frac{-z}{z - (-1 + i)} \right) + \frac{1}{2}.$$

This can also be written as

$$u(z) = \frac{2}{\pi} \cdot \tan^{-1} \left[\frac{-y - x}{-x(x + 1) - y(y - 1)} \right] + \frac{1}{2}.$$

12.3.2   Electrostatics on a Disc

EXAMPLE 10 We now analyze the problem that was set up in part II of §12.2. We do so by conformally mapping the unit disc (in the z plane) to the upper halfplane (in the w plane) by way of the mapping

$$w = f(z) = i \cdot \frac{1 - z}{1 + z}.$$

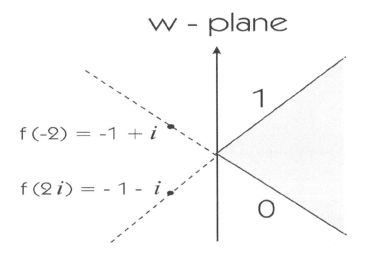

Figure 12.6: The angular region in the w-plane that is the image of the lens-shaped region in the z-plane.

See Figure 12.7. Observe that this conformal mapping takes the upper half of the unit circle to the positive real axis, the lower half of the unit circle to the negative real axis, and the point 1 to the origin.

Thus we are led to consider the following boundary value problem on the upper halfplane in the w variable: we seek a harmonic function on the upper halfplane with boundary value 0 on the positive real axis and boundary value 1 on the negative real axis. Certainly the function

$$\varphi(w) = \frac{1}{\pi} \arg w$$

does the job, if we assume that $0 \leq \arg w < 2\pi$. We pull this solution back to the disc by way of the mapping f:

$$u = \varphi \circ f.$$

This function u is harmonic on the unit disc, has boundary value 0 on the upper half of the circle, and boundary value 1 on the lower half of the circle.

Our solution may be written more explicitly as

$$u(z) = \frac{1}{\pi} \arg \left[i \cdot \frac{1-z}{1+z} \right]$$

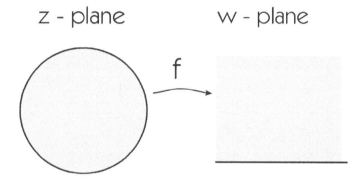

Figure 12.7: The angular region in the w-plane that is the image of the lens-shaped region in the z-plane.

or as

$$u(z) = \frac{1}{\pi} \tan^{-1}\left[\frac{1 - x^2 - y^2}{2y}\right].$$

Of course for this last form of the solution to make sense, we must take $0 \leq \arctan t \leq \pi$ and we must note that

$$\lim_{\substack{t \to 0 \\ t > 0}} \arctan t = 0 \quad \text{and} \quad \lim_{\substack{t \to 0 \\ t < 0}} \arctan t = \pi.$$

□

12.3.3   Incompressible Fluid Flow around a Post

EXAMPLE 11 We study the classic problem of the flow of an incompressible fluid around a cylindrical post.

Recall the potential function φ from the end of the discussion in III of §12.2. If

$$V = p + iq$$

is the velocity vector, then we may write

$$V = \varphi_x + i\varphi_y = \text{grad }\varphi. \tag{12.7}$$

Since φ is harmonic, we may select a conjugate harmonic function ψ (see §§3.2.2) for φ. Because of the Cauchy–Riemann equations, the velocity vector will be tangent to any curve $\psi(x, y) = $ constant. The function ψ is called the *stream function* for the flow. The curves $\psi(x, y) = $ constant are called *streamlines* of the fluid flow. We call the holomorphic function

$$H(x + iy) = \varphi(x, y) + i\psi(x, y) \qquad (12.8)$$

the *complex potential* of the fluid flow.

Using the Cauchy–Riemann equations twice, we can write $H'(z)$ as

$$H'(z) = \varphi_x(x, y) + i\psi_x(x, y)$$

or

$$H'(z) = \varphi_x(x, y) - i\varphi_y(x, y).$$

Thus formula (12.7) for the velocity becomes

$$V = \overline{H'(z)}. \qquad (12.9)$$

As a result,

$$\text{speed} = |V| = |\overline{H'(z)}| = |H'(z)|.$$

The analysis we have just described means that, in order to solve an incompressible fluid flow problem, we need to find the complex potential function H.

Now consider an incompressible fluid flow with a circular obstacle as depicted in Figure 12.8. The flow is from left to right. Far away from the obstacle, the flow is very nearly along horizontal lines parallel to the x-axis. But near to the obstacle the flow will be diverted. Our job is to determine analytically just how that diversion takes place.

We consider the circular obstacle to be given by the equation $x^2 + y^2 = 1$. Elementary symmetry considerations allow us to restrict attention to the flow in the upper halfplane. See Figure 12.9.

The boundary of the region W of the flow (Figure 12.9) is mapped to the boundary of the upper halfplane U in the w variable (this boundary is just the u-axis) by the conformal mapping

$$w = f(z) = z + \frac{1}{z}. \qquad (12.10)$$

And the region itself (shaded in Figure 12.10) is mapped to the upper halfplane in the w variable. The complex potential for a uniform flow in the upper halfplane of the w variable is

$$G(w) = Aw, \qquad (12.11)$$

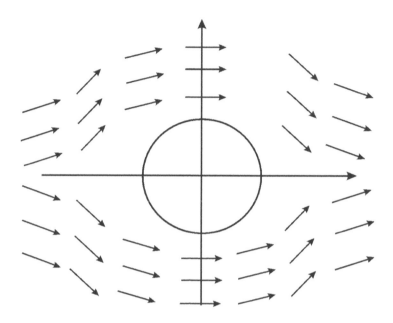

Figure 12.8: Incompressible fluid flow with a circular obstacle.

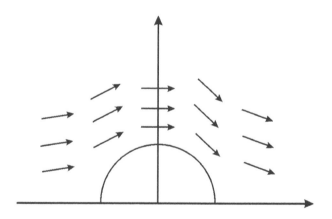

Figure 12.9: Restriction of attention to the flow in the upper halfplane.

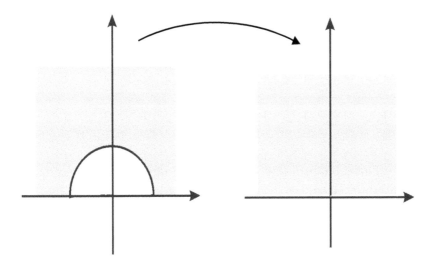

Figure 12.10: Conformal mapping of the region of the flow to the upper halfplane.

where A is a positive constant. Composing this potential with the mapping f, we find that the corresponding potential on W is

$$H(z) = G \circ f(z) = A \cdot \left(z + \frac{1}{z} \right). \tag{12.12}$$

The velocity [referring to equation (12.9)] is then

$$V = \overline{H'(z)} = A \left(1 - \frac{1}{\overline{z}^2} \right). \tag{12.13}$$

Note that V approaches A as $|z|$ increases monotonically to infinity. We conclude that the flow is almost uniform, and parallel to the x-axis, at points that are far from the circular obstacle. Finally observe that $V(\overline{z}) = \overline{V(z)}$, so that, by symmetry, formula (12.13) also represents the velocity of the flow in the lower halfplane.

According to equation (12.8), the stream function for our problem, written in polar coordinates, is just the imaginary part of H, or

$$\psi = A \left(r - \frac{1}{r} \right) \sin \theta. \tag{12.14}$$

The streamlines

$$A \left(r - \frac{1}{r} \right) \sin \theta = C \tag{12.15}$$

are symmetric to the y-axis and have asymptotes parallel to the x-axis. When $C = 0$, then the streamline consists of the circle $r = 1$ and the parts of the x axis that lie outside the unit circle.   □

Exercises

1. Solve the Dirichlet problem on the first quadrant of the complex plane with boundary data
$$\varphi(z) = \begin{cases} 1 & \text{if} & z = iy, \ y > 0 \\ 0 & \text{if} & z = x, \ x > 0. \end{cases}$$

2. Solve the Dirichlet problem on the upper halfplane with boundary data
$$\varphi(x + i0) = \sin x.$$

3. Show that if the velocity potential is $\phi = A \log r$ (for A a positive constant) for the flow in the region $\{re^{i\theta} : r \geq r_0\}$, then the streamlines are the halflines $\{re^{i\theta} : \theta = c, r \geq r_0\}$. Also show that the rate of flow outward through each complete circle about the origin is $2\pi A$. This result corresponds to a source of that strength at the origin.

4. Write the complex potential for the flow around a cylinder $r = r_0$ when the velocity V at a point z approaches a real constant A as the point recedes from the cylinder.

5. Consider a flow in the first quadrant that comes in downward, parallel to the y-axis, then at the origin turns and flows outward to the right in parallel with the x-axis. At what point of the first quadrant is the fluid pressure greatest?

6. At an interior point of a region of flow, the fluid pressure cannot be less than the pressure at all points in a neighborhood. Provide a justification for this statement.

7. Use the function $\log z$ to find an expression for the bounded, steady-state temperature in a plate having the form of the first quadrant. Assume that the edges are perfectly insulated and that the edges have

temperatures $T(x, 0) = 0$ and $T(0, y) = 1$. Find the isotherms and lines of flow, and draw some of them.

8. Find the expression for temperatures $T(r, \theta)$ in a semicircular plate $\{re^{i\theta} : 0 \leq r \leq 1, 0 \leq \theta \leq \pi\}$ with insulated faces assuming that $T = 1$ along the radial edge $\{re^{i\theta} : \theta = 0, 0 < r < 1\}$ and $T = 0$ along the rest of the boundary.

9. Consider a thin plate having insulated faces whose shape is the upper half of the region in the plane enclosed by an ellipse with foci at $(\pm 1, 0)$. The temperature on the elliptical part of the boundary is $T = 1$. The temperature along the segment $-1 < x < 1$ in the x-axis is $T = 0$. The remainder of the boundary along the x-axis is assumed to be insulated. Use a conformal mapping to a rectangle to find the heat flow lines.

10. Explain the maximum principle for harmonic functions in terms of heat.

12.4 Numerical Techniques of Conformal Mapping

In practical applications, computer techniques for calculating conformal mappings have been shown to be decisive. For instance, it is a standard technique to conformally map the complement of an airfoil (Figure 12.11) to the complement of a circle in order to study a boundary value problem on the complement. Of course the boundary of the airfoil is not given by any standard geometric curve (circle, parabola, ellipse, etc.), and the only hope of getting accurate information about the conformal mapping is by means of numerical analysis.

The literature on numerical techniques of conformal mapping is extensive. The reference [PAS] is a gateway to some of the standard references. Here we present only a brief sketch of some of the ideas.

12.4.1 Numerical Approximation of the Schwarz–Christoffel Mapping

Let P be a polygon in the complex plane with vertices $w_1, \ldots, w_n$. We wish to conformally map, by way of a mapping g, the upper halfplane

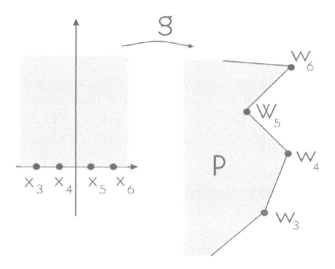

Figure 12.11: An airfoil.

U to the interior of P—see Figure 12.12. The vertices $w_1, \ldots, w_{n-1}$ in the boundary P will have pre-images $x_1, \ldots, x_{n-1}$ under g in ∂U. These latter points are called *pre-vertices*. It is standard to take $\pm\infty \in \partial U$ to be the pre-images of the last vertex w_n. Recall that, associated to each corner w_j, we have a "right-turn angle" θ_j. See Figure 12.13.

A conformal self-map of the unit disc is completely determined once the images of three boundary points are known. Since the disc and the upper halfplane are conformally equivalent by way of the Cayley transform (§§11.3), it follows that, in specifying the Schwarz–Christoffel map, we may select three of the x_j's arbitrarily.

We will take $x_1 = -1, x_2 = 0$, and $x_n = +\infty$. It remains to determine the other $n-3$ vertices (which may *not* be freely chosen). The Schwarz–Christoffel map has the form

$$ A \oint d\zeta + B \tag{12.16} $$

(see Figure 12.13). The choice of A will determine the size of the image and the choice of B will determine the position. Thus we need to choose $x_3, \ldots, x_{n-1}$ so that the image mapping has the right *shape*.

Specifying that the image of the Schwarz–Christoffel map g has the pre-specified shape (that is, the shape of P) is equivalent to demanding that

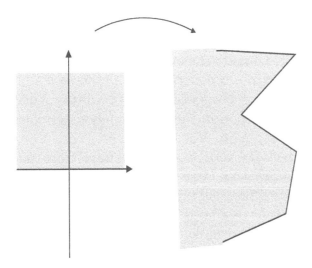

Figure 12.12: Mapping the upper halfplane to a polygonal region.

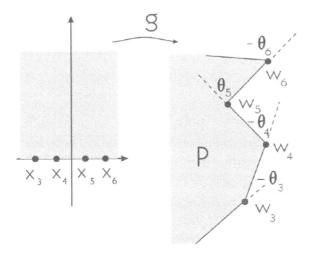

Figure 12.13: Vertices, pre-vertices, and right-turn angles.

$$\frac{|g(x_j) - g(x_{j-1})|}{|g(x_2) - g(x_1)|} = \frac{|w_j - w_{j-1}|}{|w_2 - w_1|} , \qquad j = 3, 4, \ldots, n - 1. \qquad (12.17)$$

The theory of Schwarz–Christoffel guarantees that the angles in the image will be correct, and the $n - 3$ equations (12.17) force $n - 2$ of the side lengths to have the correct proportions. With a little planar geometry, one sees that this information in turn completely determines the shape of the image of the mapping g.

Finding the correct pre-vertices for a Schwarz–Christoffel mapping problem is called the *Schwarz–Christoffel parameter problem*. The problem is resolved by solving the constrained system (12.17) of $n-3$ nonlinear equations. The constraint is that $0 < x_3 < x_4 < \cdots < x_{n-1} < \infty$. Unfortunately, standard numerical solution techniques (such as Newton's method and its variants) do not allow for constraints such as these. We can eliminate the constraint with a change of variable: Set

$$\widetilde{x}_j = \log(x_j - x_{j-1}) , \qquad j = 3, 4, \ldots, n - 1. \qquad (12.18)$$

The $\tilde{x}_j$'s are arbitrary real numbers, with no constraints. The new system, expressed in terms of the variables $\widetilde{x}_3, \widetilde{x}_4, \ldots, \widetilde{x}_{n-1}$, can be solved using a suitable version of Newton's method.

Once we have solved the Schwarz–Christoffel parameter problem, then we wish to calculate the actual conformal map. This entails calculating the Schwarz–Christoffel integral

$$\oint_0^z (\zeta - x_1)^{\theta_1/\pi} (\zeta - x_2)^{\theta_2/\pi} \cdots (\zeta - x_{n-1})^{\theta_{n-1}/\pi} \, d\zeta. \qquad (12.19)$$

Even when P is a very simple polygon such as a triangle, we cannot expect to evaluate the integral (12.19) by hand. There are well-known numerical techniques for evaluating an integral $\int_a^b \varphi(t) \, dt$. Let us describe three of them: Fix a partition $a = t_0 \le t_1 \le \cdots \le t_k = b$ of the interval of integration, which is such that each interval in the partition has the same length. Let $\Delta t = (b - a)/k$ denote that common length.

I. The Midpoint Rule   Set $p_j = a + (j - 1/2)\Delta t$, $j = 1, \ldots, k$. Then

$$\int_a^b \varphi(t) \, dt \approx \sum_{j=1}^k \varphi(p_j)\Delta t. \qquad (12.20)$$

If φ is smooth, then the error in this calculation is of size k^{-2}.

II. The Trapezoid Rule   In this methodology we take the points p_j at which φ is evaluated to be the interval endpoints. We have

$$\int_a^b \varphi(t)\,dt \approx \frac{\Delta t}{2}\left[\varphi(t_0) + 2\varphi(t_1) + 2\varphi(t_2) + \cdots + 2\varphi(t_{k-1}) + \varphi(t_k)\right].$$
$$(12.21)$$

If φ is smooth, then the error in this calculation is of size k^{-3}.

III. Simpson's Rule   In this methodology we use both the interval endpoints and the midpoints. We have

$$\int_a^b \varphi(t)\,dt \;\approx\; \frac{\Delta t}{6}\left\{\varphi(t_0) + \varphi(t_k) + 2\left[\varphi(t_1) + \cdots \varphi(t_{k-1})\right]\right.$$
$$\left. + 4\left[\varphi\left(\frac{t_0 + t_1}{2}\right) + \cdots + \varphi\left(\frac{t_{k-1} + t_k}{2}\right)\right]\right\}. (12.22)$$

If φ is smooth, then the error in this calculation is of size k^{-4}.

More sophisticated techniques, such as the *Newton–Cotes formulas* and *Gaussian quadrature* give even more accurate approximations.

Unfortunately, the integrand in the Schwarz–Christoffel integral has singularities at the pre-vertices x_j. Thus we must use a variant of the above-described numerical integration techniques with the partition points and with weights chosen so as to compensate for the singularities. These ideas are encapsulated in the method of *Gauss–Jacobi quadrature*.

Finally, it is often the case that the pre-vertices are very close together. This extreme proximity can work against the compensating properties of Gauss–Jacobi quadrature. The method of *compound Gauss–Jacobi quadrature* mandates that difficult intervals be heavily subdivided near the endpoints. This method results in a successful calculation of the Schwarz–Christoffel mapping.

12.4.2   Numerical Approximation to a Mapping onto

a Smooth Domain

One can construct a numerical approximation to a conformal mapping of the upper halfplane U onto a smoothly bounded domain Ω with boundary curve C by thinking of C as a polygon with infinitely many corners w_j, each with infinitesimal turning angle θ_j.

From the Schwarz–Christoffel formula for the mapping g that we discussed in §§12.4.1, we know that

$$
\begin{aligned}
g'(z) &= A(z - x_1)^{\theta_1/\pi}(z - x_2)^{\theta_2/\pi}(z - x_{n-1})^{\theta_{n-1}/\pi} \\
&= A \exp\left[\frac{1}{\pi}\sum_{j=1}^{n-1}\theta_j\log(z - x_j)\right].
\end{aligned}
\tag{12.23}
$$

As $n \to +\infty$, it is natural to think of the sum as converging to an integral. So we have derived the formula

$$
g'(z) = A \exp\left[\frac{1}{\pi}\int_{-\infty}^{\infty}\theta(x)\log(z - x)\,dx\right].
\tag{12.24}
$$

Here θ is a function that describes the turning of the curve C per unit length along the x-axis. Integrating, we find that

$$
g(z) = A\int_0^z \exp\left[\frac{1}{\pi}\int_{-\infty}^{\infty}\theta(x)\log(z - x)\,dx\right] + B.
\tag{12.25}
$$

Of course the discussion here has only been a sketch of some of the key ideas associated with numerical conformal mapping. The reference [SASN, pp. 430-443] gives a more discursive discussion, with detailed examples. The reference [SASN] also offers a further guide to the literature.

Exercises

1. Use the Schwarz–Christoffel transformation to discover the mapping

$$
w = z^m
$$

which maps the upper halfplane onto the wedge $\{w \in \mathbb{C} : |w| \geq 0, 0 \leq \arg w \leq m\pi\}$ and maps the point $z = 1$ to the point $w = 1$.

2. Derive a Schwarz–Christoffel transformation mapping the upper half-plane onto the triangle with vertices 0, i, and 1.

3. Find a Schwarz–Christoffel transformation which maps the upper half-plane onto the semi-infinite strip $\{w \in \mathbb{C} : |\operatorname{Re} w| < 1, \operatorname{Im} w > 0\}$.

4. Map the upper halfplane onto the region

$$U = \{w \in \mathbb{C} : \operatorname{Im} w < 0, \operatorname{Re} w > 0\} \cup \{w \in \mathbb{C} : 0 < \operatorname{Im} w < 1\}.$$

5. Use the Schwarz–Christoffel transformation to derive the usual conformal mapping of the upper halfplane onto the first quadrant.

6. Map the upper halfplane onto the semi-infinite strip

$$U = \{w \in \mathbb{C} : \operatorname{Re} w > 0, 0 < \operatorname{Im} w < 1\}.$$

7. Show that the transformation

$$w = \int_0^\pi \frac{d\zeta}{(1 - \zeta^2)^{2/3}}$$

maps the upper halfplane onto the interior of an equilateral triangle.

8. Map the upper halfplane onto the exterior of the semi-infinite strip

$$U = \{w \in \mathbb{C} : |\operatorname{Re} w| < 1, \operatorname{Im} w > 0\}.$$

9. Map the upper halfplane onto the region

$$U = \{w \in \mathbb{C} : \operatorname{Re} w < 0\} \cup \{w \in \mathbb{C} : \operatorname{Im} w > 0, \operatorname{Re} w < 1\}.$$

10. Map the upper halfplane onto the strip

$$U = \{w \in \mathbb{C} : 0 < \operatorname{Im} w < 1\}.$$

Chapter 13

Harmonic Functions

13.1 Basic Properties of Harmonic Functions

13.1.1 The Laplace Equation

Let F be a holomorphic function on an open set $U \subseteq \mathbb{C}$. Write $F = u+iv$, where u and v are real-valued. The real part u satisfies a certain partial differential equation known as *Laplace's equation*:

$$\left(\frac{\partial^2}{\partial x^2} + \frac{\partial^2}{\partial y^2} \right) u = 0. \tag{13.1}$$

(Of course the imaginary part v satisfies the same equation.) The verification is immediate: We know that

$$\frac{\partial}{\partial \bar{z}} F = 0$$

hence

$$\frac{\partial}{\partial z} \frac{\partial}{\partial \bar{z}} F = 0 \,.$$

Writing out this last equation and multiplying through by 4 gives

$$\left(\frac{\partial^2}{\partial x^2} + \frac{\partial^2}{\partial y^2} \right) F = 0 \,.$$

Finally, breaking up this last identity into real and imaginary parts, we see that

$$\left(\frac{\partial^2}{\partial x^2} + \frac{\partial^2}{\partial y^2} \right) u + i \left(\frac{\partial^2}{\partial x^2} + \frac{\partial^2}{\partial y^2} \right) v = 0 \,.$$

The only possible conclusion is that

$$\left(\frac{\partial^2}{\partial x^2} + \frac{\partial^2}{\partial y^2} \right) u \equiv 0 \qquad \text{and} \left(\frac{\partial^2}{\partial x^2} + \frac{\partial^2}{\partial y^2} \right) v \equiv 0 \,.$$

In this chapter we shall study systematically those C^2 functions that satisfy this equation. They are called *harmonic* functions. (Note that we encountered some of these ideas already in §§3.2.1.)

13.1.2   Definition of Harmonic Function

Recall the precise definition of harmonic function:

A real-valued function $u : U \to \mathbb{R}$ on an open set $U \subseteq \mathbb{C}$ is *harmonic* if it is C^2 on U and

$$\Delta u \equiv 0, \tag{13.2}$$

where the Laplacian Δu is defined by

$$\Delta u = \left(\frac{\partial^2}{\partial x^2} + \frac{\partial^2}{\partial y^2} \right) u. \tag{13.3}$$

13.1.3   Real- and Complex-Valued Harmonic Functions

The definition of harmonic function just given applies as well to complex-valued functions. A complex-valued function is harmonic if and only if its real and imaginary parts are each harmonic.

The first thing that we need to check is that real-valued harmonic functions are just those functions that arise as the real parts of holomorphic functions—at least locally.

13.1.4 Harmonic Functions as the Real Parts of Holomorphic Functions

If $u : D(P, r) \to \mathbb{R}$ is a harmonic function on a disc $D(P, r)$, then there is a holomorphic function $F : D(P, r) \to \mathbb{C}$ such that $\operatorname{Re} F \equiv u$ on $D(P, r)$. Let us write $F = u + iv$.

In fact v is determined by the Cauchy–Riemann equations:

$$\frac{\partial v}{\partial x} = -\frac{\partial u}{\partial y}$$
$$\frac{\partial v}{\partial y} = \frac{\partial u}{\partial x}.$$

We know from multivariable calculus that such a pair of equations can be solved—on a simply connected domain—precisely when the data functions (the functions on the right) satisfy a set of compatibility conditions:

$$\frac{\partial}{\partial y}\left(-\frac{\partial u}{\partial y}\right) = \frac{\partial}{\partial x}\left(\frac{\partial u}{\partial x}\right),$$

in other words,

$$\left(\frac{\partial^2}{\partial x^2} + \frac{\partial^2}{\partial y^2}\right) u = 0.$$

Of course we know in advance that u is harmonic, so the compatibility condition is satisfied. And therefore v will exist.

Note that v is uniquely determined by u except for an additive constant: the Cauchy–Riemann equations determine the partial derivatives of v and hence determine v up to an additive constant. One can also think of the determination, up to a constant, of v by u in another way: If $\tilde{v}$ is another function such that $u + i\tilde{v}$ is holomorphic, then $H \equiv i(v - \tilde{v}) = (u + iv) - (u + i\tilde{v})$ is a holomorphic function with zero real part; hence its image is not open. Thus H must be a constant, and v and $\tilde{v}$ differ by a constant. Any (harmonic) function v such that $u + iv$ is holomorphic is called a *harmonic conjugate* of u (again see §§3.2.2).

THEOREM 5 *If U is a simply connected open set (see §§4.2.2) and if $u : U \to \mathbb{R}$ is a harmonic function, then there is a C^2 (indeed a C^∞) harmonic function v such that $u + iv : U \to \mathbb{C}$ is holomorphic.*

Another important relationship between harmonic and holomorphic functions is this:

> If $u : U \to \mathbb{R}$ is harmonic and if $H : V \to U$ is holomorphic, then $u \circ H$ is harmonic on V.

One verifies this last assertion by simply differentiating $u \circ H$—using the chain rule. We leave the details to the interested reader.

13.1.5 Smoothness of Harmonic Functions

If $u : U \to \mathbb{R}$ is a harmonic function on an open set $U \subseteq \mathbb{C}$, then $u \in C^\infty$. In fact a harmonic function is always real analytic (has a local power series expansion in powers of x and y). This follows, for instance, because a harmonic function is locally the real part of a holomorphic function (see §§3.2.2, 13.1.4).

Exercises

1. Suppose that u_1 and u_2 have the same harmonic conjugate. Show that u_1 and u_2 differ by a constant.

2. Suppose that h is a holomorphic function on a domain U and that the real part of h is constant. What does that tell you about h?

3. Let u be a harmonic function on a domain U and $u \equiv 0$ on a nonempty open subset $V \subseteq U$. What does that tell you about u?

4. Let $u(x, y) = x^2 - y^2$. Verify that u is harmonic. Now find a harmonic function v on the unit disc so that $u + iv$ is holomorphic.

5. Let $u(x, y) = e^x \cos y$. Verify that u is harmonic. Now find a harmonic function v on the unit disc so that $u + iv$ is holomorphic.

6. Let U be a domain in $\mathbb{C}$ and let $E \subseteq U$ be a nontrivial line segment. If h is a holomorphic function on U and $h\big|_E = 0$ then it follows that $h \equiv 0$. Why? But the same assertion is not true for a harmonic function. Give an example to explain why not.

7. If u is harmonic and real-valued on a domain U and u^2 is harmonic on U then show that u is constant.

8. Let u be harmonic on a domain U and suppose that $u \cdot v$ is harmonic for every harmonic function v on U. Then show that u is constant.

9. Let u be harmonic and real-valued and nonvanishing on a domain U. Let $p \geq 1$. Show that $\triangle |u|^p = p(p-1)|u|^{p-2}|\nabla u|^2$.

10. Show that if u is real-valued and harmonic on a domain $U \subseteq \mathbb{C}$ then, about each point $P \in U$, u has a power series expansion. This will not be simply a power series expansion in z alone, but rather in z and $\bar{z}$ or, equivalently, in x and y.

11. If f is a nonvanishing harmonic function on a domain U then show that $\log |u|$ is harmonic on U.

13.2 The Maximum Principle and the Mean Value Property

13.2.1 The Maximum Principle for Harmonic Functions

If $u : U \to \mathbb{R}$ is harmonic on a connected open set U and if there is a point $P_0 \in U$ with the property that $u(P_0) = \max_{z \in U} u(z)$, then u is constant on U. Compare the maximum modulus principle for holomorphic functions in §§10.1.1. We also considered this phenomenon in Exercise 7 of Section 10.1. We shall learn another way to understand this maximum principle when we study the mean value property below.

13.2.2 The Minimum Principle for Harmonic Functions

If $u : U \to \mathbb{R}$ is a harmonic function on a connected open set $U \subseteq \mathbb{C}$ and if there is a point $P_0 \in U$ such that $u(P_0) = \min_{Q \in U} u(Q)$, then u is constant on U. Compare the minimum principle for holomorphic functions in §§10.1.3.

The reader may note that the minimum principle for holomorphic functions requires an extra hypothesis (i.e., nonvanishing of the function) while that for harmonic functions does not. The difference may be explained by noting that with harmonic functions we are considering the real-valued function u, while with holomorphic functions we must restrict attention to the modulus function $|f|$.

13.2.3   The Boundary Maximum and Minimum Principles

An important and intuitively appealing consequence of the maximum principle is the following result (which is sometimes called the "boundary maximum principle"). Recall that a continuous function on a compact set assumes a maximum value (and also a minimum value). When the function is harmonic, the maximum occurs at the boundary in the following precise sense:

> Let $U \subseteq \mathbb{C}$ be a bounded domain. Let u be a continuous, real-valued function on the closure $\overline{U}$ of U that is harmonic on U. Then

$$\max_{\overline{U}} u = \max_{\partial U} u. \tag{13.4}$$

The analogous result for the minimum is:

> Let $U \subseteq \mathbb{C}$ be a bounded domain. Let u be a continuous, real-valued function on the closure $\overline{U}$ of U that is harmonic on U. Then

$$\min_{\overline{U}} u = \min_{\partial U} u. \tag{13.5}$$

Compare the analogous results for holomorphic functions in §§10.1.1, 10.1.3.

13.2.4   The Mean Value Property

Suppose that $u : U \to \mathbb{R}$ is a harmonic function on an open set $U \subseteq \mathbb{C}$ and that $\overline{D}(P, r) \subseteq U$ for some $r > 0$. Then

$$u(P) = \frac{1}{2\pi} \int_0^{2\pi} u(P + re^{i\theta}) \, d\theta. \tag{13.6}$$

To understand why this result is true, let us simplify matters by assuming (with a simple translation of coordinates) that $P = 0$. Notice that if $k > 0$ and $u(z) = z^k$ then

$$\frac{1}{2\pi} \int_0^{2\pi} u(P + re^{i\theta})\, d\theta = \frac{1}{2\pi} \int_0^{2\pi} (re^{i\theta})^k\, d\theta = r^k \frac{1}{2\pi} \int_0^{2\pi} e^{ik\theta}\, d\theta = 0 = u(0).$$

When $k = 0$ so that $u(z) \equiv 1$, we get that

$$\frac{1}{2\pi} \int_0^{2\pi} u(P + re^{i\theta})\, d\theta = 1 = u(0).$$

Thus this is the mean value property for powers of z. But any holomorphic function is a sum of powers of z, so it follows that the mean value property will hold for any holomorphic function.

Finally, any harmonic function is the real part of a holomorphic function (at least locally on a disc) so that the result for harmonic functions follows by taking real parts.

We will see in §§13.2.4 that the mean value property characterizes harmonic functions—in the sense that any continuous function that satisfies the mean value property must be harmonic.

We conclude this subsection with two alternative formulations of the mean value property (MVP). In both, u, U, P, r are as above.

First Alternative Formulation of MVP

$$u(P) = \frac{1}{\pi r^2} \iint_{D(P,r)} u(x, y)\, dx dy. \tag{13.7}$$

Second Alternative Formulation of MVP

$$u(P) = \frac{1}{2\pi r} \int_{\partial D(P,r)} u(\zeta)\, d\sigma(\zeta), \tag{13.8}$$

where $d\sigma$ is arc-length measure on $\partial D(P, r)$.

13.2.5 Boundary Uniqueness for Harmonic Functions

If $u_1 : \overline{D}(0, 1) \to \mathbb{R}$ and $u_2 : \overline{D}(0, 1) \to \mathbb{R}$ are two continuous functions, each of which is harmonic on $D(0, 1)$ and if $u_1 = u_2$ on $\partial D(0, 1) =$

$\{z : |z| = 1\}$, then $u_1 \equiv u_2$. This assertion follows from the boundary maximum principle (13.4) applied to $u_1 - u_2$. Thus, in effect, a harmonic function u on $D(0, 1)$ that extends continuously to $\overline{D}(0, 1)$ is completely determined by its values on $\overline{D}(0, 1) \setminus D(0, 1) = \partial D(0, 1)$.

Exercises

1.   Let u be any harmonic function on a domain $U \subseteq \mathbb{C}$. In general it will *not* be the case that $|u|$ is harmonic (give an example). But it will be true that, for any $\overline{D}(P, r) \subseteq U$, we have the *sub-mean value property*

$$|u(P)| \leq \frac{1}{2\pi} \int_0^{2\pi} |u(P + re^{it})| \, dt \, .$$

Demonstrate this last inequality.

2.   Verify directly, with a calculation, that the mean value property holds for the harmonic function $u(x, y) = x^2 - y^2$.

3.   Verify directly, with a calculation, that the mean value property holds for the harmonic function $u(x, y) = e^y \sin x$.

4.   Do you find it curious that the mean value property is formulated in terms of the average over circles? Could there be a mean value property over squares? The answer is *no*. Provide an example to show that the mean value property does *not* hold over the unit square.

5.   Refer to Exercise 1. If u is harmonic and real-valued then it will satisfy the mean value property. In general we cannot expect that u^2 will satisfy the mean value property—after all, in general, u^2 will not be harmonic. But u^2 *will* satisfy a sub-mean value property. Verify this claim.

6.   Use calculus to give a verification of the First Alternative Formulation of the MVP.

7.   Use calculus to give a verification of the Second Alternative Formulation of the MVP.

8.   Show directly that if h is holomorphic and nonvanishing then $\log |h|$ satisfies the mean value property.

13.3 The Poisson Integral Formula

13.3.1 The Poisson Integral

The next result shows how to calculate a harmonic function on the disc from its "boundary values," i.e., its values on the circle that bounds the disc.

Let $u : U \to \mathbb{R}$ be a harmonic function on a neighborhood of $\overline{D}(0, 1)$. Then, for any point $a \in D(0, 1)$,

$$u(a) = \frac{1}{2\pi} \int_0^{2\pi} u(e^{it}) \cdot \frac{1 - |a|^2}{|a - e^{it}|^2} \, dt. \tag{13.9}$$

13.3.2 The Poisson Kernel

The expression

$$\frac{1}{2\pi} \frac{1 - |a|^2}{|a - e^{it}|^2}$$

is called the *Poisson kernel* for the unit disc. It is often convenient to rewrite the formula we have just enunciated by setting $a = |a|e^{i\theta} \equiv re^{i\theta}$. Then the result says that

$$u(re^{i\theta}) = \frac{1}{2\pi} \int_0^{2\pi} u(e^{it}) \frac{1 - r^2}{1 - 2r\cos(\theta - t) + r^2} \, dt.$$

In other words

$$u(re^{i\theta}) = \int_0^{2\pi} u(e^{it}) P_r(\theta - t) \, dt,$$

where

$$P_r(\theta - t) = \frac{1}{2\pi} \frac{1 - r^2}{1 - 2r\cos(\theta - t) + r^2}.$$

In fact this new integral formula follows rather naturally from results that we already know—if we simply remember to think of a harmonic function as the real part of a holomorphic function.

13.3.3 The Dirichlet Problem

The Poisson integral formula both reproduces and creates harmonic functions. But, in contrast to the holomorphic case (§§4.4), there is a simple connection between a continuous function f on $\partial D(0,1)$ and the created harmonic function u on D. The following theorem states this connection precisely. The theorem is usually called "the solution of the Dirichlet problem on the disc."

13.3.4 The Solution of the Dirichlet Problem on the Disc

THEOREM 6 (Solution of the Dirichlet Problem) *Let f be a continuous function on $\partial D(0,1)$. Define*

$$
u(z) = \begin{cases} \dfrac{1}{2\pi} \displaystyle\int_0^{2\pi} f(e^{it}) \cdot \dfrac{1 - |z|^2}{|z - e^{it}|^2} \, dt & \text{if} \quad z \in D(0,1) \\ f(z) & \text{if} \quad z \in \partial D(0,1). \end{cases}
$$

Then u is continuous on $\overline{D}(0,1)$ and harmonic on $D(0,1)$.

Closely related to this result is the *reproducing property* of the Poisson kernel:

Let u be continuous on $\overline{D}(0,1)$ and harmonic on $D(0,1)$. Then, for $z \in D(0,1)$,

$$
u(z) = \frac{1}{2\pi} \int_0^{2\pi} u(e^{it}) \cdot \frac{1 - |z|^2}{|z - e^{it}|^2} \, dt. \tag{13.10}
$$

See (13.9).

Let us begin by verifying formula (13.10). We assume for simplicity that the function u is harmonic on a neighborhood of $\overline{D}(0,1)$. Let u be as given and $z \in D(0,1)$ a fixed point. Consider the function $u \circ \varphi_{-z}$. It is still harmonic, so it satisfies the mean value property. We calculate

$$
\begin{aligned}
u \circ \varphi_{-z}(0) &= \frac{1}{2\pi} \int_0^{2\pi} u \circ \varphi_{-z}(e^{it}) \, dt \\
&= \frac{1}{2\pi i} \oint_{\partial D(0,1)} \frac{u \circ \varphi_{-z}(\zeta)}{\zeta} \, d\zeta
\end{aligned}
$$

Notice that we have transformed the real integral, which is what is usually used to express the mean value property, to a complex line integral. In doing so, we have kept in mind that $\zeta = e^{it}$ and $d\zeta = ie^{it}\, dt$. This will facilitate the change of variable that we must perform. Now we have that this last

$$
\begin{aligned}
&= \frac{1}{2\pi i} \oint_{\partial D(0,1)} \frac{u(\xi)}{\varphi_z(\xi)} \varphi'_z(\xi)\, d\xi \\
&= \frac{1}{2\pi i} \oint_{\partial D(0,1)} \frac{u(\xi)}{\frac{\xi - z}{1 - \overline{z}\xi}} \cdot \frac{1 - |z|^2}{(1 - \overline{z}\xi)^2}\, d\xi \\
&= \frac{1}{2\pi i} \oint_{\partial D(0,1)} u(\xi) \cdot \frac{1 - |z|^2}{(\xi - z)(1 - \overline{z}\xi)}\, d\xi \\
&= \frac{1}{2\pi i} \oint_{\partial D(0,1)} u(\xi) \cdot \frac{1 - |z|^2}{(\xi - z)(\overline{\xi} - \overline{z})} \cdot \overline{\xi}\, d\xi \\
&= \frac{1}{2\pi i} \oint_{\partial D(0,1)} u(\xi) \cdot \frac{1 - |z|^2}{|\xi - z|^2} \cdot \overline{\xi}\, d\xi \\
&= \frac{1}{2\pi} \int_0^{2\pi} u(e^{it}) \cdot \frac{1 - |z|^2}{|z - e^{it}|^2}\, dt \; .
\end{aligned}
$$

Here again we have interpreted the complex line integral as a real integral, using $\xi = e^{it}$, $d\xi = ie^{it}\, dt$. This is formula (13.10).

13.3.5 The Dirichlet Problem on a General Disc

A change of variables shows that the results of §§13.3.4 remain true on a general disc. To wit, let f be a continuous function on $\partial D(P, r)$. Define

$$
u(z) = \begin{cases} \dfrac{1}{2\pi} \displaystyle\int_0^{2\pi} f(P + re^{i\psi}) \cdot \dfrac{r^2 - |z - P|^2}{|(z - P) - re^{i\psi}|^2}\, d\psi & \text{if} \quad z \in D(P, r) \\[2mm] f(z) & \text{if} \quad z \in \partial D(P, r). \end{cases}
$$
(13.11)

Then u is continuous on $\overline{D}(P, r)$ and harmonic on $D(P, r)$.

If instead u is continuous on $\overline{D}(P, r)$ and harmonic on $D(P, r)$, then, for $z \in D(P, r)$,

$$
u(z) = \frac{1}{2\pi} \int_0^{2\pi} u(P + re^{i\psi}) \cdot \frac{r^2 - |z - P|^2}{|(z - P) - re^{i\psi}|^2}\, d\psi.
$$
(13.12)

Exercises

1. Explicitly calculate the Poisson integral of the function
$$f(e^{it}) = \begin{cases} 1 & \text{if} \quad 0 \le t \le \pi \\ -1 & \text{if} \quad \pi < t < 2\pi. \end{cases}$$

2. Verify by direct calculation that the Poisson kernel
$$\frac{1 - |z|^2}{|z - e^{it}|^2}$$
is harmonic as a function of z.

3. Refer to Exercise 2. The Cauchy integral formula on the unit disc says that
$$f(z) = \frac{1}{2\pi i} \oint_{\partial D(0,1)} \frac{f(\zeta)}{\zeta - z} \, d\zeta.$$
Of course, for $\zeta \in \partial D(0,1)$ fixed, the Cauchy kernel
$$C(z, \zeta) = \frac{1}{\zeta - z}$$
is holomorphic as a function of z. It turns out that the real part of
$$\frac{1}{2\pi i} \cdot \frac{1}{\zeta - z} \, d\zeta - \frac{1}{4\pi}$$
equals half of the Poisson kernel. Demonstrate this last statement. This gives another way to see that the Poisson kernel is harmonic as a function of z.

4. Derive formula (13.12) from formula (13.10).

5. It can be shown from the second law of thermodynamics (see [KRA1]) that if f is an initial distribution of heat on the boundary of a unit aluminum disc, then the solution of the Dirichlet problem on that disc (given by the Poisson integral) is the steady-state heat distribution on the disc induced by f. Use physical reasoning to draw some conclusions about the steady-state heat distribution for the f given in Exercise 1. What will be the value of the heat distribution at the origin? What will be the nature of the heat distribution in the *upper half* of the disc? What will be the nature of the heat distribution in the lower half of the disc?

6. If $f_k(e^{it}) = e^{ikt}$ for k a nonnegative integer then the solution of the Dirichlet problem on the unit disc with boundary data f_k is z^k. Demonstrate this result. If instead $f_k(e^{it}) = e^{ikt}$ for k a negative integer then the solution of the Dirichlet problem on the unit disc with boundary data f_k is $\overline{z}^{-k}$. Demonstrate this result.

7. Refer to Exercise 6. If we use the theory of Fourier series (Section 14.1) then we can express an arbitrary $f(e^{it})$ as

$$f(e^{it}) = \sum_{j=-\infty}^{\infty} a_j e^{ijt}.$$

This suggests that the corresponding solution of the Dirichlet problem will be

$$u(z) = \sum_{j=0}^{\infty} a_j z^j + \sum_{j=-\infty}^{-1} a_j (\overline{z})^{-j}.$$

Apply this philosophy to the boundary function $f(e^{it}) = \sin 2t$. Apply this philosophy to the boundary function $f(e^{it}) = \cos^2 t$.

Chapter 14

The Fourier Theory

Introductory Remarks

This chapter will sketch some connections of Fourier series, the Fourier transform, and the Laplace transform with the theory of complex variables. This will not be a tutorial in any of these three techniques. The reader who desires background should consult the delightful texts [DYM] or [KAT] or [KRA1].

The idea of Fourier series or Fourier transforms is to take a function f that one wishes to analyze and to assign to f a new function $\widehat{f}$ that contains information about the frequencies that are built into the function f. As such, the Fourier theory is a real variable theory. But complex variables can come to our aid in the calculation of, and also in the analysis of, $\widehat{f}$. It is that circle of ideas that will be explained in the present chapter.

14.1 Fourier Series

14.1.1 Basic Definitions

Fourier series takes place on the interval $[0, 2\pi)$. We think of the endpoints of this interval as being identified, so that geometrically our analysis is taking place on a circle (Figure 14.1). If f is an integrable function

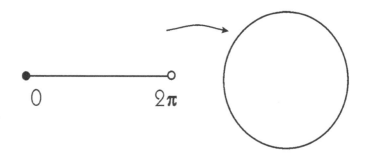

Figure 14.1: Identification of the interval with the circle.

on $[0, 2\pi)$, then we *define*[1]

$$\widehat{f}(n) = \frac{1}{2\pi} \int_0^{2\pi} f(t)e^{-int}\, dt. \tag{14.1}$$

The *Fourier series* of f is the formal expression

$$Sf(t) \sim \sum_{n=-\infty}^{\infty} \widehat{f}(n)e^{int}. \tag{14.2}$$

We call this a "formal expression" because we do not know a priori whether this series converges in any sense, and if it does converge, whether its limit is the original function f.

There is a highly developed theory of the convergence of Fourier series, but this is not the proper context in which to describe those results. Let us simply formulate one of the most transparent and useful theorems.

The partial sums of the Fourier series Sf are defined to be

$$S_N f(t) \equiv \sum_{n=-N}^{N} \widehat{f}(n)e^{int}. \tag{14.3}$$

We say that the Fourier series *converges* to the function f at the point t if

$$\lim_{N \to \infty} S_N f(t) = f(t). \tag{14.4}$$

---

[1]Already, in this particular version of the definition of the Fourier coefficients, we see complex variables coming into play. We should note that many treatments (see, for example, [SIK]) define coefficients $a_0 = [1/(2\pi)] \int_{-\pi}^{\pi} f(t)\, dt$, $a_n = [1/\pi] \int_{-\pi}^{\pi} f(t) \cos nt\, dt$, $b_n = [1/\pi] \int_{-\pi}^{\pi} f(t) \sin nt\, dt$ for $n \geq 1$. In this way one avoids the use of complex variables, but gives up something in terms of elegance and conciseness.

> **Theorem:** Let f be an integrable function on $[0, 2\pi)$. If t_0 is a point of differentiability of f, then the Fourier series Sf converges to f at t_0.

Many functions that we encounter in practice are piecewise differentiable, so this is a theorem that is straightforward to apply. In fact, if f is continuously differentiable on a compact interval I, then the Fourier series converges absolutely and uniformly to the original function f. In this respect Fourier series are much more attractive than Taylor series; for the Taylor series of even a C^∞ function f typically does not converge, and even when it does converge, it typically does not converge to f.

14.1.2   A Remark on Intervals of Arbitrary Length

It is sometimes convenient to let the interval $[-\pi, \pi]$ be the setting for our study of Fourier series. Since we think of the function f as being 2π-periodic, this results in no change in the notation or in the theory.

In applications, one often wants to do Fourier series analysis on an interval $[-L/2, L/2]$. In this setting the notation is adjusted as follows: For a function f that is integrable on $[-L/2, L/2]$, we define

$$\widehat{f}(n) = \frac{1}{L} \int_{-L/2}^{L/2} f(t) e^{-in2\pi/L} \, dt \qquad (14.5)$$

and set the Fourier series of f equal to

$$Sf(t) \sim \sum_{n=-\infty}^{\infty} \widehat{f}(n) e^{in2\pi t/L}. \qquad (14.6)$$

We shall say no more about Fourier analysis on $[-L/2, L/2]$ at this time.

14.1.3   Calculating Fourier Coefficients

The key to using complex analysis for the purpose of computing Fourier series is to note that, when $n \geq 0$, the function $\varphi_n(t) = e^{int}$ is the "boundary function" of the holomorphic function z^n. What does this mean?

We identify the interval $[0, 2\pi)$ with the unit circle S in the complex plane by way of the map

$$
\begin{aligned}
M : [0, 2\pi) &\longrightarrow S \\
t &\longmapsto e^{it}.
\end{aligned}
\tag{14.7}
$$

Of course, the circle S is the boundary of the unit disc D. If we let z be a complex variable, then, when $|z| = 1$, we know that z has the form $z = e^{it}$ for some real number t between 0 and 2π. Thus the holomorphic (analytic) function z^n, $n \geq 0$, takes the value $(e^{it})^n = e^{int}$ on the circle. By the same token, when $n < 0$, then the meromorphic function z^n takes the value $(e^{it})^n = e^{int}$ on the circle. In this way we associate, in a formal fashion, the meromorphic function

$$
F(z) = \sum_{n=-\infty}^{\infty} \widehat{f}(n) z^n
\tag{14.8}
$$

with the Fourier series

$$
Sf(t) \sim \sum_{n=-\infty}^{\infty} \widehat{f}(n) e^{int}.
\tag{14.9}
$$

This association is computationally useful, as the next example shows.

14.1.4   Calculating Fourier Coefficients Using Complex Analysis

Let us calculate the Fourier coefficients of the function $f(t) = e^{2i \sin t}$ using complex variable theory. We first recall that

$$
\sin t = \frac{1}{2i} \left[e^{it} - e^{-it} \right].
$$

Thus, using the ideas from §§14.1.3, we associate to $2i \sin t$ the analytic function

$$
z - \frac{1}{z}.
$$

As a result, we associate to f the analytic function

$$
F(z) = e^{z - 1/z} = e^z \cdot e^{-z^{-1}}.
$$

But the function on the right is easy to expand in a series:

$$F(z) = e^z \cdot e^{-z^{-1}}$$

$$= \left[\sum_{k=0}^{\infty} \frac{z^k}{k!} \right] \cdot \left[\sum_{\ell=0}^{\infty} \frac{(-1)^\ell z^{-\ell}}{\ell!} \right].$$

By the theory of the Cauchy product of series (see [KRA2]), two convergent power series may be multiplied together in just the same way as two polynomials: we multiply term by term and then gather together the resulting terms with the same power of z. We therefore find that

$$F(z) = \sum_{n=-\infty}^{\infty} z^n \left[\sum_{m=n}^{\infty} \frac{1}{m!} \frac{(-1)^n}{(m-n)!} \right].$$

In conclusion, we see that the Fourier series of our original function f is

$$Sf(x) \sim \sum_{-\infty}^{\infty} \widehat{f}(n) e^{inx}$$

with

$$\widehat{f}(n) = \sum_{m=n}^{\infty} \frac{1}{m!} \frac{(-1)^n}{(m-n)!}. \qquad \square$$

14.1.5 Steady-State Heat Distribution

The next example will harken back to our discussion of heat diffusion in §§12.2.2. But we will now use some ideas from Fourier series and from Laurent series.

EXAMPLE 12 Suppose that a thin metal heat-conducting plate is in the shape of a round disc and has radius 1. Imagine that the upper half of the circular boundary of the plate is held at constant temperature $10°$ and the lower half of the circular boundary is held at constant temperature $0°$. Describe the steady-state heat distribution on the entire plate. $\square$

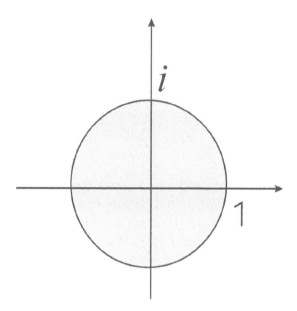

Figure 14.2: Mathematical model of the disc.

Solution: Model the disc with the interior of the unit circle in the complex plane (Figure 14.2). Identifying $[0, 2\pi)$ with the unit circle as in §§14.1.3, we are led to consider the function

$$f(t) = \begin{cases} 10 & \text{if } 0 \le t \le \pi \\ 0 & \text{if } \pi \le t < 2\pi. \end{cases}$$

Then

$$\widehat{f}(0) = \frac{1}{2\pi} \int_0^\pi 10 \, dt + \frac{1}{2\pi} \int_\pi^{2\pi} 0 \, dt = 5$$

and, for $n \ne 0$,

$$\widehat{f}(n) = \frac{1}{2\pi} \int_0^\pi 10 \cdot e^{-int} \, dt + \frac{1}{2\pi} \int_\pi^{2\pi} 0 \cdot e^{-int} \, dt = \frac{1}{2\pi} \frac{10}{in} \left[1 - e^{-in\pi} \right].$$

As a result,

$$Sf \sim 5 + \sum_{-\infty}^{-1} \frac{1}{2\pi} \cdot \frac{10}{in} \left[1 - e^{-in\pi}\right] e^{int} + \sum_{n=1}^{\infty} \frac{1}{2\pi} \cdot \frac{10}{in} \left[1 - e^{-in\pi}\right] e^{int}$$

$$= 5 + \sum_{n=1}^{\infty} \frac{1}{2\pi} \cdot \frac{10}{-in} \left[1 - e^{in\pi}\right] e^{-int} + \sum_{n=1}^{\infty} \frac{1}{2\pi} \cdot \frac{10}{in} \left[1 - e^{-in\pi}\right] e^{int}$$

$$= 5 + 2\mathrm{Re}\left(\sum_{n=1}^{\infty} \frac{1}{2\pi} \cdot \frac{10}{in} \left[1 - e^{-in\pi}\right]\right) e^{int}$$

Of course the expression in brackets is 0 when n is even. So we can rewrite our formula for the Fourier series as

$$Sf \sim 5 + \frac{20}{\pi}\mathrm{Re}\left(\frac{1}{i}\sum_{k=0}^{\infty}\frac{1}{2k+1}e^{i(2k+1)t}\right).$$

This series is associated, just as we discussed in §§14.1.3, with the analytic function

$$F(z) = 5 + \frac{20}{\pi}\mathrm{Re}\left(\frac{1}{i}\sum_{k=0}^{\infty}\frac{1}{2k+1}z^{2k+1}\right). \qquad (14.10)$$

To sum the series

$$\sum_{k=0}^{\infty}\frac{1}{2k+1}z^{2k+1},$$

we write it as

$$\sum_{k=0}^{\infty}\int_{0}^{z}\zeta^{2k}\,d\zeta = \int_{0}^{z}\left[\sum_{k=0}^{\infty}\zeta^{2k}\right]d\zeta,$$

where we have used the fact that integrals and convergent power series commute. But

$$\frac{1}{1-\alpha} = 1 + \alpha + \alpha^2 + \cdots$$

is a familiar power series expansion. Using this series with $\alpha = z^2$ we find that

$$\sum_{k=0}^{\infty}\frac{1}{2k+1}z^{2k+1} = \int_{0}^{z}\frac{1}{1-\zeta^2}\,d\zeta = \frac{1}{2}\log\left(\frac{1+z}{1-z}\right).$$

Putting this information into (14.10) yields

$$
\begin{aligned}
F(z) &= 5 + \frac{10}{\pi}\mathrm{Re}\left[\frac{1}{i}\log\left(\frac{1+z}{1-z}\right)\right] \\
&= 5 + \frac{10}{\pi}\arg\left(\frac{1+z}{1-z}\right).
\end{aligned}
$$

This function $F(z) = F(re^{i\theta})$ is the harmonic function on the disc with boundary function f. It is therefore the solution to our heat diffusion problem.

14.1.6   The Derivative and Fourier Series

Now we show how complex variables can be used to discover important formulas about Fourier coefficients. In this subsection we concentrate on the derivative.

Let f be a C^1 function on $[0, 2\pi)$. Assume that $f(0) = \lim_{t\to 2\pi-} f(t)$ and that $f'(0) = \lim_{t\to 2\pi-} f'(t)$, so that the values of f and its derivative match up at the endpoints. We want to consider how the Fourier series of f relates to the Fourier series of f'. We proceed formally.

We write

$$
f(t) \sim \sum_{n=-\infty}^{\infty} \widehat{f}(n)e^{int}
$$

and hence

$$
F(z) = \sum_{n=-\infty}^{\infty} \widehat{f}(n)z^n.
$$

A convergent power series may be differentiated termwise, so we have

$$
F'(z) = \frac{dF}{dz}(z) = \sum_{n=-\infty}^{\infty} n\widehat{f}(n)z^{n-1}.
$$

But, with $z = e^{it}$, we have

$$
\frac{dz}{dt} = \frac{d}{dt}e^{it} = ie^{it} = iz,
$$

so the chain rule tells us that

$$\frac{dF}{dt} = \frac{dF}{dz} \cdot \frac{dz}{dt}$$

$$= \sum_{n=-\infty}^{\infty} n\widehat{f}(n) z^{n-1} \cdot (iz)$$

$$= \sum_{n=-\infty}^{\infty} in\widehat{f}(n) z^{n}$$

$$= \sum_{n=-\infty}^{\infty} in\widehat{f}(n) e^{int}.$$

We conclude that the Fourier series for $f'(t)$ is

$$\sum_{n=-\infty}^{\infty} in\widehat{f}(n) e^{int}, \tag{14.11}$$

and that the Fourier coefficients for $f'(t)$ are

$$[f']\widehat{\,}(n) = in\widehat{f}(n).$$

$\square$

Exercises

1. Find the Fourier series of the function

$$f(x) = \begin{cases} \pi & \text{if } -\pi \leq x \leq \dfrac{\pi}{2} \\ 0 & \text{if } \dfrac{\pi}{2} < x \leq \pi. \end{cases}$$

2. Find the Fourier series for the function

$$f(x) = \begin{cases} 0 & \text{if } -\pi \leq x < 0 \\ 1 & \text{if } 0 \leq x \leq \frac{\pi}{2} \\ 0 & \text{if } \dfrac{\pi}{2} < x \leq \pi. \end{cases}$$

3. Find the Fourier series of the function

$$f(x) = \begin{cases} 0 & \text{if} \quad -\pi \le x < 0 \\ \sin x & \text{if} \quad 0 \le x \le \pi. \end{cases}$$

4. Solve Exercise **3** with $\sin x$ replaced by $\cos x$.

5. Find the Fourier series for each of these functions. Pay special attention to the reasoning used to establish your conclusions; consider alternative lines of thought.

(a) $f(x) = \pi$, $\quad -\pi \le x \le \pi$

(b) $f(x) = \sin x$, $\quad -\pi \le x \le \pi$

(c) $f(x) = \cos x$, $\quad -\pi \le x \le \pi$

(d) $f(x) = \pi + \sin x + \cos x$, $\quad -\pi \le x \le \pi$

Solve Exercises 6 and 7 without actually calculating the Fourier coefficients.

6. Find the Fourier series for the function given by

(a)

$$f(x) = \begin{cases} -a & \text{if} \quad -\pi \le x < 0 \\ a & \text{if} \quad 0 \le x \le \pi \end{cases}$$

for a a positive real number.

(b)

$$f(x) = \begin{cases} -1 & \text{if} \quad -\pi \le x < 0 \\ 1 & \text{if} \quad 0 \le x \le \pi \end{cases}$$

(c)

$$f(x) = \begin{cases} -\frac{\pi}{4} & \text{if} \quad -\pi \le x < 0 \\ \frac{\pi}{4} & \text{if} \quad 0 \le x \le \pi \end{cases}$$

(d)

$$f(x) = \begin{cases} -1 & \text{if} \quad -\pi \le x < 0 \\ 2 & \text{if} \quad 0 \le x \le \pi \end{cases}$$

(e)

$$f(x) = \begin{cases} 1 & \text{if} \quad -\pi \le x < 0 \\ 2 & \text{if} \quad 0 \le x \le \pi \end{cases}$$

7. Find the Fourier series for the periodic function defined by

$$f(x) = \begin{cases} -\pi & \text{if } -\pi \leq x < 0 \\ x & \text{if } 0 \leq x < \pi \end{cases}$$

Sketch the graph of the sum of this series on the interval $-5\pi \leq x \leq 5\pi$ and find what numerical sums are implied by the convergence behavior at the points of discontinuity $x = 0$ and $x = \pi$.

8. (a) Show that the Fourier series for the periodic function

$$f(x) = \begin{cases} 0 & \text{if } -\pi \leq x < 0 \\ x^2 & \text{if } 0 \leq x < \pi \end{cases}$$

is

$$f(x) = \frac{\pi^2}{6} + 2\sum_{j=1}^{\infty}(-1)^j \frac{\cos jx}{j^2}$$

$$+ \pi\sum_{j=1}^{\infty}(-1)^{j+1}\frac{\sin jx}{j} - \frac{4}{\pi}\sum_{j=1}^{\infty}\frac{\sin(2j-1)x}{(2j-1)^3}.$$

(b) Sketch the graph of the sum of this series on the interval $-5\pi \leq x \leq 5\pi$.

(c) Use the series in part **(a)** with $x = 0$ and $x = \pi$ to obtain the two sums

$$1 - \frac{1}{2^2} + \frac{1}{3^2} - \frac{1}{4^2} + - \cdots = \frac{\pi^2}{12}$$

and

$$1 + \frac{1}{2^2} + \frac{1}{3^2} + \frac{1}{4^2} + \cdots = \frac{\pi^2}{6}.$$

(d) Derive the second sum in **(c)** from the first.   **Hint:** Add $2\sum_j(1/[2j])^2$ to both sides.

* **9. (a)** Find the Fourier series for the periodic function defined by $f(x) = e^x$, $-\pi \leq x \leq \pi$. **Hint:** Recall that $\sinh x = (e^x - e^{-x})/2$.

(b) Sketch the graph of the sum of this series on the interval $-5\pi \leq x \leq 5\pi$.

(c) Use the series in (a) to establish the sums

$$\sum_{j=1}^{\infty} \frac{1}{j^2 + 1} = \frac{1}{2}\left(\frac{\pi}{\tanh \pi} - 1\right)$$

and

$$\sum_{j=1}^{\infty} \frac{(-1)^j}{j^2 + 1} = \frac{1}{2}\left(\frac{\pi}{\sinh \pi} - 1\right).$$

10. Determine whether each of the following functions is even, odd, or neither:

$$x^5 \sin x \ , \ \ x^2 \sin 2x \ , \ \ e^x \ , \ \ (\sin x)^3 \ , \ \ \sin x^2 \ ,$$

$$\cos(x + x^3) \ , \ \ x + x^2 + x^3 \ , \ \ \ln \frac{1 + x}{1 - x}.$$

11. Show that any function f defined on a symmetrically placed interval can be written as the sum of an even function and an odd function. [**Hint:** $f(x) = \frac{1}{2}[f(x) + f(-x)] + \frac{1}{2}[f(x) - f(-x)]$.]

12. Find the Fourier series for the function of period 2π defined by $f(x) = \cos x/2$, $-\pi \le x \le \pi$. Sketch the graph of the sum of this series on the interval $-5\pi \le x \le 5\pi$.

13. The functions $\sin^2 x$ and $\cos^2 x$ are both even. Show, without using any calculations, that the identities

$$\sin^2 x = \frac{1}{2}(1 - \cos 2x) = \frac{1}{2} - \frac{1}{2}\cos 2x$$

and

$$\cos^2 x = \frac{1}{2}(1 + \cos 2x) = \frac{1}{2} + \frac{1}{2}\cos 2x$$

are actually the Fourier series expansions of these functions.

14. Demonstrate the trigonometric identities

$$\sin^3 x = \frac{3}{4}\sin x - \frac{1}{4}\sin 3x \quad \text{and} \quad \cos^3 x = \frac{3}{4}x + \frac{1}{4}\cos 3x$$

and show briefly, without calculation, that these are the Fourier series expansions of the functions $\sin^3 x$ and $\cos^3 x$.

15. Show that

$$\frac{L}{2} - x = \frac{L}{\pi} \sum_{j=1}^{\infty} \frac{1}{j} \sin \frac{2j\pi x}{L}, \qquad 0 < x < L.$$

14.2 The Fourier Transform

The Fourier transform is the analogue on the real line of Fourier series coefficients for a function on $[0, 2\pi)$. For deep reasons (which are explained in [FOL]), the Fourier series on the bounded interval $[0, 2\pi)$ must be replaced by the continuous analogue of a sum, which is an integral. In this section we will learn what the Fourier transform is, and what the basic convergence question about the Fourier transform is. Then we will see how complex variable techniques may be used in the study of the Fourier transform.

14.2.1 Basic Definitions

The Fourier transform takes place on the real line $\mathbb{R}$. If f is an integrable function on $\mathbb{R}$, then we *define*

$$\widehat{f}(\xi) = \int f(x)e^{-2\pi it \cdot \xi} \, dt. \qquad (14.12)$$

The variable t is called the "space variable" and the variable ξ is called the "Fourier transform variable" (or sometimes the "phase variable"). There are many variants of this definition. Some tracts replace $-2\pi it \cdot \xi$ in the exponential with $+2\pi it \cdot \xi$. Others omit the factor of 2π. We have chosen this particular definition because it simplifies certain basic formulas in the subject. The reader who wants to learn the full story of the theory of the Fourier transform should consult [FOL] or [STW] or [KRA1].

The Fourier transform $\widehat{f}$ of an integrable function f enjoys the property that $\widehat{f}$ is continuous and vanishes at infinity. However $\widehat{f}$ itself *need not* be integrable. In fact $\widehat{f}$ can die arbitrarily slowly at infinity. This fact of life necessitates extra care in formulating results about the Fourier transform and its inverse.

Recall that we recover a function on $[0, 2\pi)$ from its sequence of Fourier coefficients by calculating a *sum*. In the theory of the Fourier transform, we recover f from $\widehat{f}$ by calculating an integral. Namely, if g is any integrable function on the real line $\mathbb{R}$, then we define

$$\overset{\vee}{g}(t) = \int g(\xi)e^{2\pi i\xi \cdot t}\, d\xi. \qquad (14.13)$$

The operation $\vee$ is called the *inverse Fourier transform*.

It turns out that, whenever the integrals in question make sense, the Fourier operations $\vee$ and $\widehat{}$ are inverse to each other. More precisely, if f is a function on the real line such that

- f is integrable,

- $\widehat{f}$ is integrable,

then

$$\overset{\vee}{\widehat{f}} = f. \qquad (14.14)$$

An easily verified hypothesis that will guarantee that both f and $\widehat{f}$ are integrable is that $f \in C^2$ and f, f', f'' be integrable.

14.2.2   Some Fourier Transform Examples that Use Complex Variables

EXAMPLE 13 Let us calculate the Fourier transform of the function

$$f(t) = \frac{1}{1 + t^2}\,. \qquad \qquad \square$$

Solution: The Fourier integral is

$$\widehat{f}(\xi) = \int_{-\infty}^{\infty} \frac{1}{1 + t^2} e^{-2\pi i t \cdot \xi}\, dt.$$

We will evaluate the integral using the calculus of residues.

For fixed ξ in $\mathbb{R}$, we thus consider the meromorphic function

$$m(z) = \frac{e^{-2\pi i z \cdot \xi}}{1 + z^2},$$

which has poles at $\pm i$.

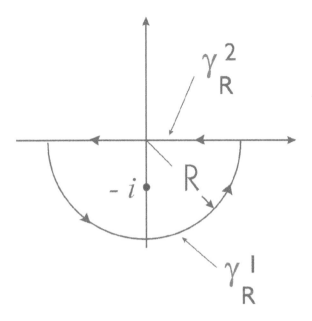

Figure 14.3: A positively oriented semicircle γ_R in the lower halfplane.

For the case $\xi > 0$ it is convenient to use as contour of integration the positively oriented semicircle γ_R of radius $R > 1$ in the lower halfplane that is shown in Figure 14.3. Of course this contour only contains the pole at $-i$. We find that

$$
\begin{aligned}
2\pi i \text{Res}_m(-i) &= \int_{\gamma_R} m(z)\, dz \\
&= \int_{\gamma_R^1} m(z)\, dz + \int_{\gamma_R^2} m(z)\, dz.
\end{aligned}
$$

The integral over γ_R^1 vanishes as $R \to +\infty$ and the integral over γ_R^2 tends to

$$
-\int_{-\infty}^{\infty} \frac{1}{1+t^2} e^{-2\pi i t \cdot \xi}\, dt.
$$

It is straightforward to calculate that

$$\text{Res}_m(-i) = \lim_{z \to -i}(z - (-i)) \cdot \frac{e^{-2\pi i \xi z}}{z^2 + 1}$$

$$= \left.\frac{e^{-2\pi i \xi z}}{z - i}\right|_{z=-i}$$

$$= \frac{e^{-2\pi i \xi(-i)}}{-2i}$$

$$= \frac{e^{-2\pi \xi}}{-2i}. \qquad (14.15)$$

Thus

$$2\pi i \text{Res}_m(-i) = -\pi e^{-2\pi \xi}.$$

We conclude that

$$\widehat{f}(\xi) = \int_{-\infty}^{\infty} \frac{1}{1 + t^2} e^{-2\pi i t \cdot \xi} \, dt = \pi e^{-2\pi \xi}.$$

A similar calculation, using the contour μ_R shown in Figure 14.4, shows that, when $\xi < 0$, then

$$\widehat{f}(\xi) = \pi e^{2\pi \xi}.$$

In summary, for any $\xi \in \mathbb{R}$,

$$\widehat{f}(\xi) = \pi e^{-2\pi |\xi|}.$$

We can now check our work using the inverse Fourier transform: We observe that both f and $\widehat{f}$ are integrable, so we calculate that

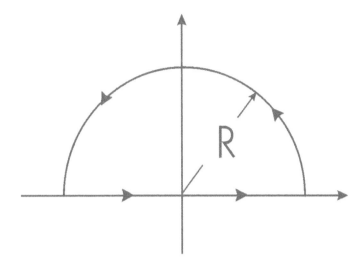

Figure 14.4: A positively oriented semicircle in the upper halfplane.

$$
\begin{aligned}
\overset{\vee}{\widehat{f}}(t) &= \int_{-\infty}^{\infty} \widehat{f} e^{2\pi i \xi \cdot t}\, d\xi \\
&= \int_{0}^{\infty} \pi e^{-2\pi \xi} e^{2\pi i \xi t}\, d\xi + \int_{-\infty}^{0} \pi e^{2\pi \xi} e^{2\pi i \xi t}\, d\xi \\
&= \int_{0}^{\infty} \pi e^{-2\pi \xi} e^{2\pi i \xi t}\, d\xi + \int_{0}^{\infty} \pi e^{-2\pi \xi} e^{-2\pi i \xi t}\, d\xi \\
&= 2\mathrm{Re}\left[\int_{0}^{\infty} \pi e^{-2\pi \xi} e^{2\pi i \xi t}\, d\xi\right] \\
&= 2\mathrm{Re}\left[\int_{0}^{\infty} \pi e^{\xi(-2\pi + 2\pi i t)}\, d\xi\right] \\
&= \mathrm{Re}\left[\frac{2\pi}{-2\pi + 2\pi i t} \cdot e^{\xi(-2\pi + 2\pi i t)}\right]_{\xi=0}^{\xi=\infty} \\
&= \mathrm{Re}\left[\frac{1}{-1 + it} \cdot (0 - 1)\right] \\
&= \frac{1}{1 + t^2}.
\end{aligned}
\tag{14.16}
$$

Observe that our calculations confirm the correctness of our Fourier transform determination. In addition, they demonstrate the validity of the Fourier inversion formula in a particular instance.

EXAMPLE 14 Physicists call a function of the form

$$f(t) = \begin{cases} \cos 2\pi t & \text{if} \quad -7/4 \le t \le 7/4 \\ 0 & \text{if} \quad t < -7/4 \text{ or } t > 7/4 \end{cases}$$

a *finite wave train.* Let us calculate the Fourier transform of this function.

Solution: The Fourier integral is

$$
\begin{aligned}
\widehat{f}(\xi) &= \int_{-\infty}^{\infty} f(t)e^{-2\pi i \xi \cdot t}\, dt \\
&= \int_{-7/4}^{7/4} (\cos 2\pi t)e^{-2\pi i \xi \cdot t}\, dt \\
&= \frac{1}{2}\left[\int_{-7/4\pi}^{7/4\pi} e^{2\pi i t}e^{-2\pi i \xi \cdot t}\, dt + \int_{-7/4\pi}^{7/4\pi} e^{-2\pi i t}e^{-2\pi i \xi \cdot t}\, dt \right] \\
&= \frac{1}{2}\left[\int_{-7/4\pi}^{7/4\pi} e^{(2\pi i - 2\pi i \xi)t}\, dt + \int_{-7/4\pi}^{7/4\pi} e^{(-2\pi i - 2\pi i \xi)t}\, dt \right] \\
&= \frac{1}{2}\left(\left[\frac{1}{2\pi i - 2\pi i \xi}e^{(2\pi i - 2\pi i \xi)t} \right]_{t=-7/4}^{t=7/4} \right. \\
&\qquad \left. + \left[\frac{1}{-2\pi i - 2\pi i \xi}e^{(-2\pi i - 2\pi i \xi)t} \right]_{t=-7/4}^{t=7/4} \right) \\
&= \frac{1}{2}\left\{ \frac{1}{2\pi i(1-\xi)}\left[e^{2\pi i(1-\xi)[7/4]} - e^{2\pi i(1-\xi)[-7/4]} \right] \right. \\
&\qquad \left. + \frac{1}{2\pi i(-1-\xi)}\left[e^{-2\pi i(1+\xi)[7/4]} - e^{-2\pi i(1+\xi)[-7/4]} \right] \right\}
\end{aligned}
$$

$$= \frac{1}{2} \left\{ \frac{1}{2\pi i (1-\xi)} 2i \sin\left(\frac{7\pi}{2}(1-\xi)\right) \right.$$

$$\left. - \frac{1}{2\pi i (-1-\xi)} 2i \sin\left(\frac{7\pi}{2}(1+\xi)\right) \right\}$$

$$= \frac{1}{2\pi(1-\xi)} \left[\sin\frac{7\pi}{2} \cos\frac{7\pi}{2}\xi - \cos\frac{7\pi}{2} \sin\frac{7\pi}{2}\xi \right]$$

$$- \frac{1}{2\pi(-1-\xi)} \left[\sin\frac{7\pi}{2} \cos\frac{7\pi}{2}\xi + \cos\frac{7\pi}{2} \sin\frac{7\pi}{2}\xi \right]$$

$$= \frac{1}{2\pi(1-\xi)} \left(-\cos\frac{7\pi}{2}\xi \right) + \frac{1}{2\pi(1+\xi)} \left(-\cos\frac{7\pi}{2}\xi \right)$$

$$= \frac{1}{2\pi} \cos\frac{7\pi}{2}\xi \left\{ \frac{-1}{1+\xi} - \frac{1}{1-\xi} \right\}$$

$$= -\frac{1}{\pi} \left(\cos\frac{7\pi}{2}\xi \right) \frac{1}{1-\xi^2}.$$

In summary,

$$\widehat{f}(\xi) = -\frac{1}{\pi} \cos\left(\frac{7\pi}{2}\xi\right) \frac{1}{1-\xi^2}.$$

We may now perform a calculation to confirm the Fourier inversion formula in this example. The calculus of residues will be seen to be a useful tool in the process.

Now

$$\overset{\vee}{\widehat{f}}(t) = -\frac{1}{\pi} \int_{-\infty}^{\infty} \frac{1}{1-\xi^2} \cos\frac{7\pi}{2}\xi \cdot e^{2\pi i \xi t} \, d\xi$$

$$= -\frac{1}{2\pi} \int_{-\infty}^{\infty} \frac{1}{1-\xi^2} \left[e^{(7\pi/2)\xi i} + e^{-(7\pi/2)\xi i} \right] e^{2\pi i \xi t} \, d\xi$$

$$= -\frac{1}{2\pi} \int_{-\infty}^{\infty} \frac{1}{1-\xi^2} e^{i\xi[(7\pi/2)+2\pi t]} \, d\xi$$

$$- \frac{1}{2\pi} \int_{-\infty}^{\infty} \frac{1}{1-\xi^2} e^{i\xi[(-7\pi/2)+2\pi t]} \, d\xi$$

$$\equiv I + II. \tag{14.17}$$

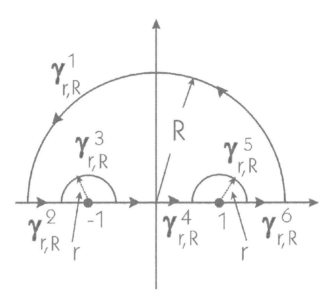

Figure 14.5: The curve $\gamma_{r,R}$.

First we analyze expression I. For $t \geq -7/4$ fixed, the expression $(7\pi/2) + 2\pi t$ is non-negative. Thus the exponential expression will be bounded (i.e., the real part of the exponent will be non-positive), if we integrate the meromorphic function

$$\frac{1}{1-z^2} e^{iz[(7\pi/2)+2\pi t]}$$

on the curve $\gamma_{r,R}$ exhibited in Figure 14.5. It is easy to see that the integral over $\gamma_{r,R}^1$ tends to zero as $R \to +\infty$. And the integrals over $\gamma_{r,R}^2, \gamma_{r,R}^4, \gamma_{r,R}^6$ tend to the integral that is I as $r \to 0$. It remains to evaluate the integrals over $\gamma_{r,R}^3$ and $\gamma_{r,R}^5$. We do the first and leave the second for the reader.

Now

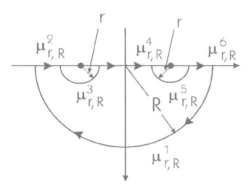

Figure 14.6: The curve $\mu_{r,R}$.

$$-\frac{1}{2\pi}\oint_{\gamma_{r,R}^3}\frac{1}{1-z^2}e^{iz[(7\pi/2)+2\pi t]}\,dz$$

$$=\quad -\frac{1}{2\pi}\int_\pi^0\frac{1}{1-(-1+re^{i\theta})^2}e^{i(-1+re^{i\theta})[(7\pi/2)+2\pi t]}ire^{i\theta}\,d\theta$$

$$=\quad -\frac{1}{2\pi}\int_\pi^0\frac{ire^{i\theta}}{2re^{i\theta}-r^2e^{2i\theta}}e^{i(-1+re^{i\theta})[(7\pi/2)+2\pi t]}\,d\theta$$

$$=\quad -\frac{i}{2\pi}\int_\pi^0\frac{1}{2-re^{i\theta}}e^{i(-1+re^{i\theta})[(7\pi/2)+2\pi t]}\,d\theta$$

$$\xrightarrow{(r\to0)}\quad -\frac{i}{2\pi}\int_\pi^0\frac{1}{2}e^{-i[(7\pi/2)+2\pi t]}\,d\theta$$

$$=\quad -\frac{1}{4}e^{-2\pi it}. \qquad\qquad (14.18)$$

A similar calculation shows that

$$-\frac{1}{2\pi}\oint_{\gamma^5_{r,R}}\frac{1}{1-z^2}e^{iz[(7\pi/2)+2\pi t]}\,dz=-\frac{1}{4}e^{2\pi it}.$$

In sum,

$$I=\frac{1}{2\pi}\oint_{\gamma^3_{r,R}}\quad+\quad\frac{1}{2\pi}\oint_{\gamma^5_{r,R}}\quad=-\frac{1}{2}\cos 2\pi t.$$

To analyze the integral II, we begin by fixing $t>7/4$. Then we will have $(-7\pi/2)+2\pi t>0$. If we again use the contour in Figure 14.5, then the exponential in the meromorphic function

$$\frac{1}{1-z^2}e^{iz[(-7\pi/2)+2\pi t]}$$

will be bounded on the curve $\gamma^1_{r,R}$. Of course the integral over $\gamma^1_{r,R}$ tends to zero as $R\to+\infty$. The integrals over $\gamma^2_{r,R},\gamma^4_{r,R},\gamma^6_{r,R}$ in sum tend to the integral that defines II. Finally, a calculation that is nearly identical to the one that we just performed for I shows that

$$-\frac{1}{2\pi}\oint_{\gamma^3_{r,R}}\frac{1}{1-z^2}e^{iz[(-7\pi/2)+2\pi t]}\,dz=\frac{1}{4}e^{-2\pi it}.$$

Similarly,

$$-\frac{1}{2\pi}\oint_{\gamma^5_{r,R}}\frac{1}{1-z^2}e^{iz[(-7\pi/2)+2\pi t]}\,dz=\frac{1}{4}e^{-2\pi it}.$$

Therefore

$$II=\frac{1}{2}\cos 2\pi t.$$

In summary, we see that on the common domain $t>7/4$ we have

$$I+II=-\frac{1}{2}\cos 2\pi t+\frac{1}{2}\cos 2\pi t=0.$$

This value agrees with $f(t)$ when $t>7/4$.

Similar calculations for $t<-7/4$ (but using the contour shown in Figure 14.6) show that

$$\overset{\vee}{\widehat{f}}(t)=0.$$

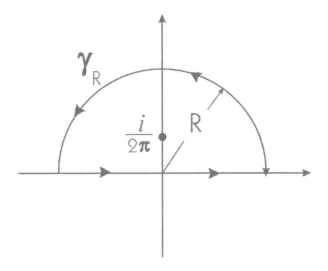

Figure 14.7: The contour used to analyze the integral.

The remaining, and most interesting, calculation is for $-7/4 \le t \le 7/4$. We have already calculated I for that range of t. To calculate II, we use the contour in Figure 14.7. The result is that

$$II = -\frac{1}{2}\cos 2\pi t.$$

Therefore

$$\overset{\vee}{\hat{f}}(t) = -(I + II) = \frac{1}{2}\cos 2\pi t + \frac{1}{2}\cos 2\pi t = \cos 2\pi t. \qquad (14.19)$$

This confirms Fourier inversion for the finite wave train.

14.2.3 Solving a Differential Equation Using the Fourier Transform

Suppose that $f \in C^1(\mathbb{R})$ and f and that both f and f' are integrable. Then

$$
\begin{aligned}
\widehat{f'}(\xi) &= \int_{-\infty}^{\infty} f'(t) e^{-2\pi i t \xi}\, dt \\
&= f(t) e^{-2\pi i t}\Big|_{-\infty}^{\infty} + 2\pi i \xi \int_{-\infty}^{\infty} f(t) e^{-2\pi i t \xi}\, dt\,.
\end{aligned}
$$

The fact that f, f' are integrable guarantees that the boundary term vanishes. We conclude that

$$\widehat{f'}(\xi) = 2\pi i \xi \widehat{f}(\xi)\,. \tag{14.20}$$

This formula is elementary but important. It is analogous to formula (14.11) for Fourier series coefficients. We can use it to solve a differential equation:

EXAMPLE 15 Use the Fourier transform to solve the differential equation

$$f''(t) - f(t) = \varphi(t), \tag{14.21}$$

where

$$\varphi(t) = \begin{cases} e^{-t} & \text{if } t > 0 \\ 0 & \text{if } t \le 0. \end{cases}$$

Solution: We begin by applying the Fourier transform to both sides of equation (14.21). The result is

$$-4\pi^2 \xi^2 \widehat{f}(\xi) - \widehat{f}(\xi) = \widehat{\varphi}.$$

An easy calculation shows that

$$\widehat{\varphi}(\xi) = -\frac{1}{1 + 2\pi i \xi}.$$

Thus, under the Fourier transform, our ordinary differential equation has become

$$-4\pi^2 \xi^2 \widehat{f}(\xi) - \widehat{f}(\xi) = -\frac{1}{1 + 2\pi i \xi}$$

or

$$\widehat{f}(\xi) = \frac{1}{(4\pi^2\xi^2 + 1)(1 + 2\pi i \xi)}. \tag{14.22}$$

Of course the expression on the right-hand side of (14.22) has no singularities on the real line (thanks to complex variables) and is integrable. So we may apply the Fourier inversion formula (§§14.2.1) to both sides of (14.22) to obtain

$$f(t) = \overset{\vee}{\widehat{f}}(t) = \left(\frac{1}{(4\pi^2\xi^2 + 1)(1 + 2\pi i \xi)}\right)^{\vee}. \tag{14.23}$$

We can find the function f if we can evaluate the expression on the right-hand side of (14.23). This amounts to calculating the integral

$$\int_{-\infty}^{\infty} \frac{1}{(4\pi^2\xi^2 + 1)(1 + 2\pi i \xi)} e^{2\pi i \xi t} \, d\xi.$$

We will do so for $t > 0$ (the most interesting set of values for t, given the data function φ in the differential equation) and let the reader worry about $t \leq 0$.

It is helpful to use the calculus of residues to evaluate the integral in (14.2.3). We use the contour in Figure 14.7, chosen (with $R >> 1$) so that the exponential in the integrand will be bounded when the variable is on the curve and $t > 0$. The pole of

$$m(z) = \frac{1}{(4\pi^2 z^2 + 1)(1 + 2\pi i z)} e^{2\pi i z t} = \frac{1}{(1 - 2\pi i z)(1 + 2\pi i z)^2} e^{2\pi i z t}$$

that lies inside of γ_R is at $P = i/[2\pi]$. This is a pole of order two. We use the formula in §8.1. The result is that

$$\mathrm{Res}_m(P) = \frac{-i}{4\pi} \left[\frac{1}{2} e^{-t} + t e^{-t}\right].$$

As usual, the value of the integral

$$\int_{-\infty}^{\infty} \frac{1}{(4\pi^2\xi^2 + 1)(1 + 2\pi i \xi)} e^{2\pi i \xi t} \, d\xi$$

is

$$(-2\pi i) \cdot \frac{-i}{4\pi} \left[\frac{1}{2} e^{-t} + t e^{-t}\right] = -\frac{1}{2} \left[\frac{1}{2} e^{-t} + t e^{-t}\right].$$

This is the solution f (at least when $t > 0$) of our differential equation (14.21).

Exercises

1. Use the Fourier transform to solve the differential equation

$$f''(t) + 2f(t) = \varphi(t),$$

where
$$\varphi(t) = \begin{cases} e^{-4t} & \text{if} \quad t > 1 \\ 0 & \text{if} \quad t \le 1. \end{cases}$$

2. Find the Fourier transform of the function

$$\varphi(t) = \begin{cases} \cos t & \text{if} \quad |t| \le 4\pi \\ 0 & \text{if} \quad |t| > 4\pi. \end{cases}$$

3. Refer to Exercise 2. Use the Fourier transform to solve the differential equation
$$f''(t) + 2f'(t) + 2f(t) = \varphi(t),$$

where
$$\varphi(t) = \begin{cases} \cos t & \text{if} \quad |t| \le 4\pi \\ 0 & \text{if} \quad |t| > 4\pi. \end{cases}$$

4. Find the Fourier transform of $f(t) = e^{-|t|}$.

5. Find the Fourier transform of $g(t) = e^{-t^2}$.

6. Use residue theory to calculate the Fourier transform of $f(t) = 1/(t^4 + 1)$. Also confirm the inversion formula in this case.

7. Use the Fourier transform to solve the differential equation

$$f''(t) + 4f'(t) + f(t) = \varphi(t),$$

where
$$\varphi(t) = \begin{cases} 0 & \text{if} \quad t < 0 \\ e^{-t} & \text{if} \quad t \ge 0. \end{cases}$$

8. Use the Fourier transform to solve the differential equation

$$f''(t) + f'(t) + f(t) = e^{-t^2}.$$

9. Let $\mathcal{F}$ denote the Fourier transform. Show that $\mathcal{F}^4 = \mathrm{id}$, where id is the identity operator. What conclusion can you draw about the eigenvalues of the Fourier transform?

10. Show that if f is an integrable function whose Fourier transform is identically 0, then $f \equiv 0$. What does this result tell you about the univalence of the Fourier transform?

Chapter 15

Other Transforms

15.1   The Laplace Transform

15.1.1   Prologue

Let f be an integrable function on the half-line $\{t \in \mathbb{R} : t \geq 0\}$. [We implicitly assume that $f(t) = 0$ when $t < 0$.]

In many contexts, it is convenient to think of the Fourier transform

$$\widehat{f}(\xi) = \int f(t)e^{-2\pi i \xi \cdot t}\, dt \tag{15.1}$$

as a function of the complex variable ξ. In fact when $\operatorname{Im}\xi < 0$ and $t > 0$ the exponent in the integrand has negative real part so the exponential is bounded and the integral converges. For suitable f one can verify, using Morera's theorem (§4.2) for instance, that $\widehat{f}(\xi)$ is a *holomorphic* function of ξ. It is particularly convenient to let ξ be pure imaginary: the customary notation is $\xi = -is/(2\pi)$ for $s \geq 0$. Then we have defined a new function

$$F(s) = \int f(t)e^{-st}\, dt. \tag{15.2}$$

We call F the *Laplace transform* of f. Sometimes, instead of writing F, we write $\mathcal{L}(f)$.

The Laplace transform is a useful tool because

- It has formal similarities to the Fourier transform.

- It can be applied to a larger class of functions than the Fourier transform (since the factor e^{-st} decays rapidly at infinity).

- It is often straightforward to compute.

The lesson here is that it is sometimes useful to modify a familiar mathematical operation (in this case the Fourier transform) by letting the variable be complex (in this case producing the Laplace transform).

We now provide just one example to illustrate the utility of the Laplace transform.

15.1.2   Solving a Differential Equation Using the Laplace Transform

EXAMPLE 16 Use the Laplace transform to solve the ordinary differential equation

$$f''(t) + 3f'(t) + 2f(t) = \sin t \qquad (15.3)$$

subject to the initial conditions $f(0) = 1$, $f'(0) = 0$.

Solution: Working by analogy with Example 15, we calculate the Laplace transform of both sides of the equation. Integrating by parts (as we did when studying the Fourier transform—§§14.2.3), we can see that

$$\mathcal{L}(f')(s) = s \cdot \mathcal{L}f(s) - 1$$

and

$$\mathcal{L}(f'')(s) = s^2 \mathcal{L}f(s) - s.$$

(These formulas are correct when all the relevant integrals converge. See also the Table of Laplace Transforms in the back of the book.)

Thus equation (15.3) is transformed to

$$[s^2 \mathcal{L}f(s) - s] + 3[s \cdot \mathcal{L}f(s) - 1] + 2\mathcal{L}f(s) = \big(\mathcal{L}[\sin t]\big)(s). \qquad (15.4)$$

A straightforward calculation (using either integration by parts or complex variable methods) shows that

$$\mathcal{L}[\sin t](s) = \frac{1}{s^2 + 1}.$$

So equation (15.4) becomes

$$[s^2 \mathcal{L}f(s) - s] + 3[s \cdot \mathcal{L}f(s) - 1] + 2\mathcal{L}f(s) = \frac{1}{s^2 + 1}.$$

Just as with the Fourier transform, the Laplace transform has transformed the differential equation to an algebraic equation. We find that

$$\mathcal{L}f(s) = \frac{1}{s^2 + 3s + 2} \cdot \left[\frac{1}{s^2 + 1} + s + 3 \right].$$

We use the method of partial fractions to break up the right-hand side into simpler components. The result is

$$\mathcal{L}f(s) = \frac{5/2}{s + 1} + \frac{-6/5}{s + 2} + \frac{-3s/10 + 1/10}{s^2 + 1}.$$

Now our job is to find the inverse Laplace transform of each expression on the right. One way to do this is by using the Laplace inversion formula

$$f(t) = \int_{-i\infty}^{\infty} F(s)e^{st}\, ds.$$

However, the most common method is to use a Table of Laplace Transforms, as in [ZWI, pp. 559–564] or the Table of Laplace Transforms in the back of this book. From such a table, we find that

$$f(t) = \frac{5}{2}e^{-t} - \frac{6}{5}e^{-2t} - \frac{3}{10}\cos t + \frac{1}{10}\sin t.$$

The reader may check that this is indeed a solution to the differential equation (15.3).

Exercises

1. Calculate the Laplace transform of $f(t) = t^m$ for m a positive integer.

2. Calculate the Laplace transform of $f(t) = \sin t$.

3. Calculate the Laplace transform of $g(t) = e^{mt}$ for m a real number.

4. Solve the differential equation

$$f''(t) + 2f'(t) + f(t) = \cos t \,,$$

subject to the conditions $f(0) = 2$, $f'(0) = 0$, using the Laplace transform.

5. Solve the differential equation

$$f''(t) - f'(t) + 3f(t) = e^t \,,$$

subject to the conditions $f(0) = 1$, $f'(0) = 1$, using the Laplace transform.

6. Solve the differential equation

$$f''(t) + f'(t) + 4f(t) = t^2 \,,$$

subject to the conditions $f(0) = 1$, $f'(0) = 0$, using the Laplace transform.

7. Calculate the inverse Laplace transform of

$$\varphi(s) = \frac{1}{s(s^2 + 1)} \,.$$

8. Calculate the inverse Laplace transform of

$$\varphi(s) = \frac{s}{(s^2 + 4)(s + 2)} \,.$$

9. Calculate the inverse Laplace transform of

$$\varphi(s) = \frac{s + 1}{(s + 1)^2 + 4} \,.$$

10. Calculate the inverse Laplace transform of

$$\varphi(s) = \frac{1}{s^4} \,.$$

15.2 The z-Transform

The z-transform, under that particular name, is more familiar in the engineering community than in the mathematics community. Mathematicians group this circle of ideas with the notion of generating function and with allied ideas from finite and combinatorial mathematics (see, for instance [STA]). Here we give a quick introduction to the z-transform and its uses.

15.2.1 Basic Definitions

Let $\{a_j\}_{j=-\infty}^{+\infty}$ be a doubly infinite sequence. The z-transform of this sequence is defined to be the series

$$A(z) = \sum_{j=-\infty}^{\infty} a_j z^{-j}. \tag{15.5}$$

If this series converges on some annulus centered at the origin, then of course it defines a holomorphic function. Often the properties of the original sequence $\{a_j\}_{j=-\infty}^{+\infty}$ can be studied by way of the holomorphic function A.

The reference [ZWI, pp. 231, 543] explains the relationship between the z-transform and other transforms that we have discussed. The reference [HEN, v. 2, pp. 322, 327, 332, 334, 335, 336, 350] gives further instances of the technique of the z-transform.

15.2.2 Population Growth by Means of the z-Transform

We present a typical example of the use of z-transform.

EXAMPLE 17 During a period of growth, a population of salmon has the following two properties:
(15.6) The population, on average, reproduces at the rate of 3% per month.
(15.7) One hundred new salmon swim upstream and join the population

each month.

If $a(n)$ is the population in month n, then find a formula for $a(n)$.

Solution: Let P denote the initial population. Then we may describe the sequence in this recursive manner:

$$\begin{aligned}
a(0) &= P \\
a(1) &= a(0) \cdot (1 + .03) + 100 \\
a(2) &= a(1) \cdot (1 + .03) + 100
\end{aligned}$$

$$\text{etc.}$$

Because we are going to use the theory of the z-transform, it is convenient to postulate that $a(n) = 0$ for $n < 0$.

Let us assume that $\{a(n)\}$ has a z-transform $A(z)$—at least when z is sufficiently large. It is also convenient to think of each part of the recursion as depending on n. So let us set

$$P(n) = \begin{cases} P & \text{if } n = -1 \\ 0 & \text{if } n \neq -1 \end{cases}$$

and

$$s(n) = \begin{cases} 100 & \text{if } n \geq 0 \\ 0 & \text{if } n < 0. \end{cases}$$

Then our recursion can be expressed as

$$a(n+1) = 1.03a(n) + P(n) + s(n).$$

We multiply both sides of this equation by z^{-n} and sum over n to obtain

$$\sum_n a(n+1)z^{-n} = 1.03 \sum_n a(n)z^{-n} + \sum_n P(n)z^{-n} + \sum_n s(n)z^{-n}$$

or

$$z \cdot A(z) = 1.03A(z) + Pz + \frac{100z}{z-1}.$$

Here, for the last term, we have used the elementary fact that

$$\sum_{n=0}^{\infty} z^{-n} = \sum_{n=0}^{\infty} (1/z)^n = \frac{1}{1 - 1/z} = \frac{z}{z-1},$$

valid for $|z| > 1$.

Rearranging equation (15.2.2), we find that

$$A(z) = \frac{Pz^2 + (100 - P)z}{(z-1)(z-1.03)} = z \cdot \frac{Pz + (100 - P)}{(z-1)(z-1.03)} ,$$

valid for $|z|$ sufficiently large.

Of course we may decompose this last expression for $A(z)$ into a partial fractions decomposition:

$$A(z) = z \cdot \left[\frac{P + 100/.03}{z - 1.03} - \frac{100/.03}{z - 1} \right].$$

We rewrite the terms in preparation of making a Laurent expansion:

$$A(z) = \left(P + \frac{100}{.03} \right) \cdot \left[\frac{1}{1 - 1.03/z} \right] - \frac{100}{.03} \cdot \frac{1}{1 - 1/z}.$$

For $|z| > 1.03$, we may use the standard expansion

$$\frac{1}{1 - \alpha} = \sum_{n=0}^{\infty} \alpha^n , \qquad |\alpha| < 1$$

to obtain

$$A(z) = \left(P + \frac{100}{.03} \right) \sum_{n=0}^{\infty} (1.03)^n z^{-n} - \frac{100}{.03} \sum_{n=0}^{\infty} z^{-n}.$$

Note that we have obtained the expansion of $A(z)$ as a z-series! Its coefficients must therefore be the $a(n)$. We conclude that

$$a(n) = \begin{cases} \left(P + \frac{100}{.03} \right) (1.03)^n - \frac{100}{.03} & \text{if} \quad n \geq 0 \\ 0 & \text{if} \quad n < 0. \end{cases}$$

It is easy to see that this problem could have been solved without the aid of the z-transform. But the z-transform was a useful device for keeping track of information.

Exercises

1. Show that the z-transform is a linear operator. That is to say, if $A(z)$ is the z-transform of $\{a(j)\}$ and $B(z)$ is the z-transform of $\{b(j)\}$, then $\alpha A(z) + \beta B(z)$ is the z-transform of $\{\alpha a(j) + \beta b(j)\}$ for any real constants α, and β.

2. Let $A(z)$ be the z-transform of $\{a(j)\}$. Show that the z-transform of the weighted sequence $\{ja(j)\}$ is $-zA'(z)$. Show that the annulus of convergence is the same for both z-transforms.

3. If $A(z)$ is the z-transform of the sequence $\{a(j)\}$, then show that the z-transform of the exponentially weighted sequence $\{\alpha^j a(j)\}$ (for $\alpha > 0$) is $A(z/\alpha)$.

4. Verify the following z-transforms:

   (a) $a(j) = \begin{cases} 1 & \text{if} & j = 0 \\ 0 & \text{if} & j > 0 \\ 0 & \text{if} & j < 0 \end{cases}$ $A(z) \equiv 1$.

   (b) $a(j) = 1$ for all j $\qquad\qquad$ $A(z) = z/(z-1)$.
   (c) $a(j) = j$ for all j $\qquad\qquad$ $A(z) = a/(z-1)^2$.
   (d) $a(j) = \alpha^j$ for $\alpha > 0$ $\qquad\quad$ $A(z) = z/(z-\alpha)$.
   (e) $a(j) = \sin j\omega$ for $\omega > 0$ $\quad$ $A(z) = (z\sin\omega)/(z^2 - 2z\cos\omega + 1)$.

5. Find the inverse z-transform for the following functions in the indicated annulus.

   (a) $A(z) = 1/(1 + 1/(3z))$ $\qquad\qquad$ $|z| < 1/3$
   (b) $A(z) = z^4/(z+2)$ $\qquad\qquad\qquad$ $|z| < 2$
   (c) $A(z) = (z+2)/(2z^2 - 7z + 3)$ $\qquad$ $1/2 < |z| < 3$
   (d) $A(z) = (1 - \alpha z)/(\alpha - 1/z)$ $\qquad$ $|z| > 1/|\alpha|$

6. Suppose that the sequence $\{a_j\}$ satisfies $a_j = 0$ for $j < 0$. Let $A(z)$ be its z-transform. Show that $\lim_{z \to \infty} A(z) = 0$.

7. Let $A(z)$ be the z-transform of the sequence $\{a_j\}_{j=-\infty}^{\infty}$, valid in the annulus $a < |z| < b$. Show that

$$a(j) = \frac{1}{2\pi i} \oint_\Gamma A(\zeta)\zeta^{j-1}\, d\zeta$$

for all j and Γ a positively oriented circle of center 0 and radius between a and b.

8. The *unilateral z-transform* is defined to be

$$A^+(z) = \sum_{j=0}^{\infty} a(j)z^j .$$

Use the unilateral z-transform to solve these difference equations.

(a) $a(j+1) + 2a(j) = 1$, $a(0) = 1$

(b) $a(j+2) - 5a(j+1) + 6a(j) = 1$, $a(0) = 2,\ a(1) = 3$

9. Refer to Exercise 8. Calculate the unilateral z-transform of the backward shift sequence $a(j - J)$ for some fixed $J > 0$.

Chapter 16

Partial Differential Equations and Boundary Value Problems

16.1 Fourier Methods in the Theory of Differential Equations

In fact an entire separate book could be written about the applications of Fourier analysis to differential equations and to other parts of mathematical analysis. The subject of Fourier series grew up hand in hand with the analytical areas to which it is applied. In the present brief section we merely indicate a couple of examples.

16.1.1 Remarks on Different Fourier Notations

In Section 14.1, we found it convenient to define the Fourier coefficients of an integrable function on the interval $[0, 2\pi]$ to be

$$\widehat{f}(n) = \frac{1}{2\pi} \int_0^{2\pi} f(x) e^{-inx} \, dx \,.$$

From the point of view of pure mathematics, this complex notation has been shown to be useful, and it has become standardized.

But, in applications, there are other Fourier paradigms. They are easily seen to be equivalent to the one we have already introduced. The reader who wants to be conversant in this subject should be aware of these different ways of writing the basic ideas of Fourier series. We will introduce one of them now, and use it in the ensuing discussion.

If f is integrable on the interval $[-\pi, \pi]$ (note that, by 2π-periodicity, this is not essentially different from $[0, 2\pi]$), then we define the Fourier coefficients

$$a_0 = \frac{1}{2\pi} \int_{-\pi}^{\pi} f(x)\, dx\,,$$

$$a_n = \frac{1}{\pi} \int_{-\pi}^{\pi} f(x) \cos nx\, dx \qquad \text{for } n \geq 1\,,$$

$$b_n = \frac{1}{\pi} \int_{-\pi}^{\pi} f(x) \sin nx\, dx \qquad \text{for } n \geq 1\,.$$

This new notation is not essentially different from the old, for

$$\widehat{f}(n) = \frac{1}{2}\left[a_n + ib_n\right]$$

for $n \geq 1$. The change in normalization (i.e., whether the constant before the integral is $1/\pi$ or $1/2\pi$) is dictated by the observation that we want to exploit the fact (so that our formulas come out in a neat and elegant fashion) that

$$\frac{1}{2\pi} \int_0^{2\pi} |e^{-int}|^2\, dt = 1\,,$$

in the theory from Section 14.1 and that

$$\frac{1}{2\pi} \int_{-\pi}^{\pi} 1^2\, dx = 1\,,$$

$$\frac{1}{\pi} \int_{-\pi}^{\pi} |\cos nt|^2\, dt = 1 \quad \text{for } n \geq 1\,,$$

$$\frac{1}{\pi} \int_{-\pi}^{\pi} |\sin nt|^2\, dt = 1 \quad \text{for } n \geq 1$$

in the theory that we are about to develop.

It is clear that any statement (as in Section 14.1) that is formulated in the language of $\widehat{f}(n)$ is easily translated into the language of a_n and b_n and vice versa. In the present discussion we shall use a_n and b_n just because that is the custom, and because it is convenient for the points that we want to make.

16.1.2 The Dirichlet Problem on the Disc

We now study the two-dimensional Laplace equation, which is

$$\triangle = \frac{\partial^2 u}{\partial x^2} + \frac{\partial^2 u}{\partial y^2} = 0. \qquad (16.1)$$

This is probably the most important differential equation of mathematical physics. It describes a steady-state heat distribution, electrical fields, and many other important phenomena of nature.

It will be useful for us to write this equation in polar coordinates. To do so, recall that

$$r^2 = x^2 + y^2 \;, \quad x = r\cos\theta \;, \quad y = r\sin\theta \,.$$

Thus

$$\frac{\partial}{\partial r} = \frac{\partial x}{\partial r}\frac{\partial}{\partial x} + \frac{\partial y}{\partial r}\frac{\partial}{\partial y} = \cos\theta\frac{\partial}{\partial x} + \sin\theta\frac{\partial}{\partial y}$$

$$\frac{\partial}{\partial \theta} = \frac{\partial x}{\partial \theta}\frac{\partial}{\partial x} + \frac{\partial y}{\partial \theta}\frac{\partial}{\partial y} = -r\sin\theta\frac{\partial}{\partial x} + r\cos\theta\frac{\partial}{\partial y}$$

We may solve these two equations for the unknowns $\partial/\partial x$ and $\partial/\partial y$. The result is

$$\frac{\partial}{\partial x} = \cos\theta\frac{\partial}{\partial r} - \frac{\sin\theta}{r}\frac{\partial}{\partial \theta} \quad \text{and} \quad \frac{\partial}{\partial y} = \sin\theta\frac{\partial}{\partial r} - \frac{\cos\theta}{r}\frac{\partial}{\partial \theta}.$$

A tedious calculation now reveals that

$$
\begin{aligned}
\triangle = \frac{\partial^2}{\partial x^2} + \frac{\partial^2}{\partial y^2} &= \left(\cos\theta\frac{\partial}{\partial r} - \frac{\sin\theta}{r}\frac{\partial}{\partial \theta}\right)\left(\cos\theta\frac{\partial}{\partial r} - \frac{\sin\theta}{r}\frac{\partial}{\partial \theta}\right) \\
&\quad + \left(\sin\theta\frac{\partial}{\partial r} - \frac{\cos\theta}{r}\frac{\partial}{\partial \theta}\right)\left(\sin\theta\frac{\partial}{\partial r} - \frac{\cos\theta}{r}\frac{\partial}{\partial \theta}\right) \\
&= \frac{\partial^2}{\partial r^2} + \frac{1}{r}\frac{\partial}{\partial r} + \frac{1}{r^2}\frac{\partial^2}{\partial \theta^2}.
\end{aligned}
$$

Let us use the so-called separation of variables method to analyze our partial differential equation (16.1). We will seek a solution $w = w(r,\theta) =$

$u(r) \cdot v(\theta)$ of the Laplace equation. Using the polar form, we find that this leads to the equation

$$u''(r) \cdot v(\theta) + \frac{1}{r} u'(r) \cdot v(\theta) + \frac{1}{r^2} u(r) \cdot v''(\theta) = 0.$$

Thus

$$\frac{r^2 u''(r) + r u'(r)}{u(r)} = -\frac{v''(\theta)}{v(\theta)}.$$

Since the left-hand side depends only on r, and the right-hand side only on θ, both sides must be constant. Denote the common constant value by λ.

Then we have

$$v'' + \lambda v = 0 \qquad\qquad (16.2)$$

and

$$r^2 u'' + r u' - \lambda u = 0. \qquad\qquad (16.3)$$

If we demand that v be continuous and periodic, then we must insist that $\lambda > 0$ and in fact that $\lambda = n^2$ for some nonnegative integer n.[1] For $n = 0$ the only suitable solution is $v \equiv$ constant and for $n > 0$ the general solution (with $\lambda = n^2$) is

$$y = A \cos n\theta + B \sin n\theta,$$

as you can verify directly.

We set $\lambda = n^2$ in equation (16.3), and obtain

$$r^2 u'' + r u' - n^2 u = 0, \qquad\qquad (16.4)$$

which is Euler's equidimensional equation. The change of variables $r = e^z$ transforms this equation to a linear equation with constant coefficients, and that can in turn be solved with standard techniques. To wit, the equation that we now have is

$$u'' - n^2 u = 0.$$

The variable is now z. We guess a solution of the form $u(z) = e^{\alpha z}$. Thus

$$\alpha^2 e^{\alpha z} - n^2 e^{\alpha z} = 0 \qquad\qquad (16.5)$$

---

[1]More explicitly, $\lambda = 0$ gives a linear function for a solution and $\lambda < 0$ gives an exponential function for a solution.

so that
$$\alpha^2 = \pm n\,.$$

Hence the solutions of (16.5) are
$$u(z) = e^{nz} \quad \text{and} \quad u(z) = e^{-nz}$$

provided that $n \neq 0$. It follows that the solutions of the original Euler equation (16.4) are
$$u(r) = r^n \quad \text{and} \quad u(r) = r^{-n} \qquad \text{for } n \neq 0\,.$$

In case $n = 0$ the solution is readily seen to be $u = 1$ or $u = \ln r$.
    The result is
$$u = A + B \ln r \qquad \text{if } n = 0\,;$$
$$u = A r^n + B r^{-n} \qquad \text{if } n = 1, 2, 3, \dots\,.$$

We are most interested in solutions u that are continuous at the origin; so we take $B = 0$ in all cases. The resulting solutions are

$$
\begin{aligned}
n &= 0\,, & w &= \text{a constant } a_0/2\,; \\
n &= 1\,, & w &= r(a_1 \cos\theta + b_1 \sin\theta)\,; \\
n &= 2\,, & w &= r^2(a_2 \cos 2\theta + b_2 \sin 2\theta)\,; \\
n &= 3\,, & w &= r^3(a_3 \cos 3\theta + b_3 \sin 3\theta)\,;
\end{aligned}
$$

$$\dots$$

Of course any finite sum of solutions of Laplace's equation is also a solution. The same is true for infinite sums. Thus we are led to consider

$$w = w(r, \theta) = \frac{1}{2}a_0 + \sum_{j=0}^{\infty} r^j (a_j \cos j\theta + b_j \sin j\theta)\,.$$

On a formal level, letting $r \to 1^-$ in this last expression gives

$$\frac{1}{2}a_0 + \sum_{j=1}^{\infty} (a_j \cos j\theta + b_j \sin j\theta)\,.$$

We draw all these ideas together with the following physical rubric. Consider a thin aluminum disc of radius 1, and imagine applying a heat

Figure 16.1: A thin aluminum disc of radius 1.

distribution to the boundary of that disc. In polar coordinates, this distribution is specified by a function $f(\theta)$. We seek to understand the steady-state heat distribution on the entire disc. See Figure 16.1. So we seek a function $w(r, \theta)$, continuous on the closure of the disc, which agrees with f on the boundary and which represents the steady-state distribution of heat inside. Some physical analysis shows that such a function w is the solution of the boundary value problem

$$
\begin{aligned}
\triangle w &= 0, \\
u\Big|_{\partial D} &= f.
\end{aligned}
$$

According to the calculations we performed prior to this last paragraph, a natural approach to this problem is to expand the given function f in its sine/cosine series:

$$
f(\theta) = \frac{1}{2}a_0 + \sum_{j=1}^{\infty}(a_j \cos j\theta + b_j \sin j\theta)
$$

and then posit that the w we seek is

$$
w(r, \theta) = \frac{1}{2}a_0 + \sum_{j=1}^{\infty} r^j (a_j \cos j\theta + b_j \sin j\theta).
$$

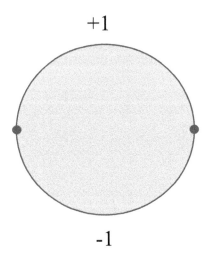

Figure 16.2: Data for the Dirichlet problem.

This process is known as *solving the Dirichlet problem on the disc with boundary data f.*

EXAMPLE 18 Let us follow the paradigm just sketched to solve the Dirichlet problem on the disc with $f(\theta) = 1$ on the top half of the boundary and $f(\theta) = -1$ on the bottom half of the boundary. See Figure 16.2.

The function f is odd. Hence the Fourier series for f will involve only the odd sine terms. This means that $a_n = 0$ for all n. We only use the b_n. It is thus straightforward to calculate that the Fourier series (sine series) expansion for this f is

$$f(\theta) = \frac{4}{\pi}\left(\sin\theta + \frac{\sin 3\theta}{3} + +\frac{\sin 5\theta}{5} + \cdots\right).$$

The solution of the Dirichlet problem is therefore

$$w(r,\theta) = \frac{4}{\pi}\left(r\sin\theta + \frac{r^3\sin 3\theta}{3} + +\frac{r^5\sin 5\theta}{5} + \cdots\right). \qquad \Box$$

16.1.3   The Poisson Integral

In the last section we have presented a formal procedure with series for solving the Dirichlet problem. But in fact it is possible to produce a closed formula for this solution. This we now do.

Referring back to our sine series expansion for f, and the resulting expansion for the solution of the Dirichlet problem, we recall for $j \geq 1$ that

$$a_j = \frac{1}{\pi} \int_{-\pi}^{\pi} f(\phi) \cos j\phi \, d\phi \quad \text{and} \quad b_j = \frac{1}{\pi} \int_{-\pi}^{\pi} f(\phi) \sin j\phi \, d\phi \,.$$

Thus

$$w(r, \theta) = \frac{1}{2} a_0 + \sum_{j=1}^{\infty} r^j \left(\frac{1}{\pi} \int_{-\pi}^{\pi} f(\phi) \cos j\phi \, d\phi \cos j\theta \right.$$

$$\left. + \frac{1}{\pi} \int_{-\pi}^{\pi} f(\phi) \sin j\phi \, d\phi \sin j\theta \right) \,.$$

This, in turn, equals

$$\frac{1}{2} a_0 + \frac{1}{\pi} \sum_{j=1}^{\infty} r^j \int_{-\pi}^{\pi} f(\phi) \left[\cos j\phi \cos j\theta + \sin j\phi \sin j\theta d\phi \right]$$

$$= \frac{1}{2} a_0 + \frac{1}{\pi} \sum_{j=1}^{\infty} r^j \int_{-\pi}^{\pi} f(\phi) \left[\cos j(\theta - \phi) d\phi \right] \,.$$

We finally simplify our expression to

$$w(r, \theta) = \frac{1}{\pi} \int_{-\pi}^{\pi} f(\phi) \left[\frac{1}{2} + \sum_{j=1}^{\infty} r^j \cos j(\theta - \phi) \right] d\phi \,.$$

It behooves us, therefore, to calculate the sum inside the integral. For simplicity, we let $\alpha = \theta - \phi$ and then we let

$$z = re^{i\alpha} = r(\cos \alpha + i \sin \alpha) \,.$$

Likewise

$$z^n = r^n e^{in\alpha} = r^n(\cos n\alpha + i \sin n\alpha) \,.$$

Let $\operatorname{Re} z$ denote the real part of the complex number z. Then

$$
\frac{1}{2} + \sum_{j=1}^{\infty} r^j \cos j\alpha \;=\; \operatorname{Re}\left[\frac{1}{2} + \sum_{j=1}^{\infty} z^j\right]
$$

$$
=\; \operatorname{Re}\left[-\frac{1}{2} + \frac{1}{1-z}\right]
$$

$$
=\; \operatorname{Re}\left[\frac{1+z}{2(1-z)}\right]
$$

$$
=\; \operatorname{Re}\left[\frac{(1+z)(1-\bar{z})}{2|1-z|^2}\right]
$$

$$
=\; \frac{1-|z|^2}{2|1-z|^2}
$$

$$
=\; \frac{1-r^2}{2(1-2r\cos\alpha + r^2)} \,.
$$

Putting the result of this calculation into our original formula for w we finally obtain the Poisson integral formula:

$$
w(r,\theta) = \frac{1}{2\pi} \int_{-\pi}^{\pi} \frac{1-r^2}{1-2r\cos\alpha + r^2} f(\phi)\, d\phi \,.
$$

Observe what this formula does for us: It expresses the solution of the Dirichlet problem with boundary data f as an explicit integral of a universal expression (called a *kernel*) against that data function f.

There is a great deal of information about w and its relation to f contained in this formula. As just one simple instance, we note that when r is set equal to 0 then we obtain

$$
w(0,\theta) = \frac{1}{2\pi} \int_{-\pi}^{\pi} f(\phi)\, d\phi \,.
$$

This says that the value of the steady-state heat distribution at the origin is just the average value of f around the circular boundary.

EXAMPLE 19 Let us use the Poisson integral formula to solve the Dirichlet problem for the boundary data $f(\phi) = e^{2i\phi}$. We know that the solution is given by

$$
\begin{aligned}
w(r, \theta) &= \frac{1}{2\pi} \int_{-\pi}^{\pi} \frac{1 - r^2}{1 - 2r \cos \alpha + r^2} f(\phi) \, d\phi \\
&= \frac{1}{2\pi} \int_{-\pi}^{\pi} \frac{1 - r^2}{1 - 2r \cos \alpha + r^2} e^{2i\phi} \, d\phi \, .
\end{aligned}
$$

With some effort, one can evaluate this integral to find that

$$
w(r, \theta) = r^2 e^{2i\theta} \, .
$$

In complex notation, w is the function $z \mapsto z^2$.   □

16.1.4   The Wave Equation

We consider the wave equation

$$
a^2 y_{xx} = y_{tt} \tag{16.6}
$$

on the interval $[0, \pi]$ with the boundary conditions

$$
y(0, t) = 0
$$

and

$$
y(\pi, t) = 0 \, .
$$

This equation, with boundary conditions, is a mathematical model for a vibrating string with the ends (at $x = 0$ and $x = \pi$) pinned down. The function $y(x, t)$ describes the ordinate of the point x on the string at time t. See Figure 16.3.

Physical considerations dictate that we also impose the initial conditions

$$
\left. \frac{\partial y}{\partial t} \right|_{t=0} = 0 \tag{16.7}
$$

(indicating that the initial velocity of the string is 0) and

$$
y(x, 0) = f(x) \tag{16.8}
$$

(indicating that the initial configuration of the string is the graph of the function f).

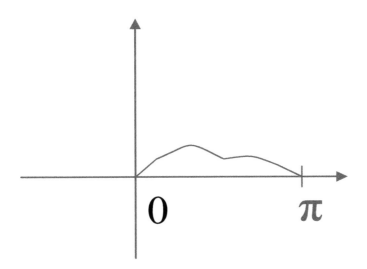

Figure 16.3: A vibrating string.

We solve the wave equation using a version of separation of variables. For convenience, we assume that the constant $a = 1$. We guess a solution of the form $u(x, t) = \alpha(x) \cdot \beta(t)$. Putting this guess into the differential equation

$$u_{xx} = u_{tt}$$

gives

$$\alpha''(x)\beta(t) = \alpha(x)\beta''(t).$$

We may obviously separate variables, in the sense that we may write

$$\frac{\alpha''(x)}{\alpha(x)} = \frac{\beta''(t)}{\beta(t)}.$$

The left-hand side depends only on x while the right-hand side depends only on t. The only way this can be true is if

$$\frac{\alpha''(x)}{\alpha(x)} = \lambda = \frac{\beta''(t)}{\beta(t)}$$

for some constant λ. But this gives rise to two second-order linear, ordinary differential equations that we can solve explicitly:

$$\alpha'' = \lambda \cdot \alpha \tag{16.9}$$

$$\beta'' = \lambda \cdot \beta. \tag{16.10}$$

Observe that this is the *same* constant λ in both of these equations. Now, as we have already discussed, we want the initial configuration of the string to pass through the points $(0,0)$ and $(\pi, 0)$. We can achieve these conditions by solving (16.9) with $\alpha(0) = 0$ and $\alpha(\pi) = 0$.

This problem has a nontrivial solution if and only if $\lambda = n^2$ for some positive integer n, and the corresponding function is

$$\alpha_n(x) = \sin nx.$$

For this same λ, the general solution of (16.10) is

$$\beta(t) = A \sin nt + B \cos nt.$$

If we impose the requirement that $\beta'(0) = 0$, so that (16.7) is satisfied, then $A = 0$ and we find the solution

$$\beta(t) = B \cos nt.$$

This means that the solution we have found of our differential equation with the given boundary and initial conditions is

$$y_n(x, t) = \sin nx \cos nt. \tag{16.11}$$

And in fact any finite sum with constant coefficients (or *linear combination*) of these solutions will also be a solution:

$$y = \alpha_1 \sin x \cos t + \alpha_2 \sin 2x \cos 2t + \cdots \alpha_k \sin kx \cos kt.$$

This is called the "principle of superposition".

Ignoring the rather delicate issue of convergence, we may claim that any *infinite* linear combination of the solutions (16.11) will also be a solution:

$$y = \sum_{j=1}^{\infty} b_j \sin jx \cos jt. \tag{16.12}$$

Now we must examine the final condition (16.8). The mandate $y(x, 0) = f(x)$ translates to

$$\sum_{j=1}^{\infty} b_j \alpha_j(x) = y(x, 0) = f(x) \tag{16.13}$$

or

$$\sum_{j=1}^{\infty} b_j \sin jx = y(x, 0) = f(x) \,. \qquad (16.14)$$

Thus we demand that f have a valid Fourier series expansion. We know from our studies earlier in this chapter that such an expansion is valid for a rather broad class of functions f. Thus the wave equation is solvable in considerable generality.

We know that our eigenfunctions α_j satisfy

$$\alpha_m'' = -m^2 \alpha_m \qquad \text{and} \qquad \alpha_n'' = -n^2 \alpha_n \,.$$

Multiply the first equation by α_n and the second by α_m and subtract. The result is

$$\alpha_n \alpha_m'' - \alpha_m \alpha_n'' = (n^2 - m^2) \alpha_n \alpha_m$$

or

$$[\alpha_n \alpha_m' - \alpha_m \alpha_n']' = (n^2 - m^2) \alpha_n \alpha_m \,.$$

We integrate both sides of this last equation from 0 to π and use the fact that $\alpha_j(0) = \alpha_j(\pi) = 0$ for every j. The result is

$$0 = [\alpha_n \alpha_m' - \alpha_m \alpha_n']\Big|_0^{\pi} = (n^2 - m^2) \int_0^{\pi} \alpha_m(x) \alpha_n(x) \, dx \,.$$

Thus

$$\int_0^{\pi} \sin mx \sin nx \, dx = 0 \qquad \text{for } n \neq m \qquad (16.15)$$

or

$$\int_0^{\pi} \alpha_m(x) \alpha_n(x) \, dx = 0 \qquad \text{for } n \neq m \,. \qquad (16.16)$$

Of course this is a standard fact from calculus. It played an important (tacit) role in Section 14.1, when we first learned about Fourier series. It is commonly referred to as an "orthogonality condition," and is fundamental to the Fourier theory and the more general Sturm–Liouville theory. We now see how the condition arises naturally from the differential equation.

In view of the orthogonality condition (16.16), it is natural to integrate both sides of (16.13) against $\alpha_k(x)$. The result is

$$
\begin{aligned}
\int_0^\pi f(x) \cdot \alpha_k(x)\, dx &= \int_0^\pi \left[\sum_{j=0}^\infty b_j \alpha_j(x) \right] \cdot \alpha_k(x)\, dx \\
&= \sum_{j=0}^\infty b_j \int_0^\pi \alpha_j(x)\alpha_k(x)\, dx \\
&= \frac{\pi}{2} b_k\,.
\end{aligned}
$$

Here we use the fact that $\int_0^\pi \alpha_j \alpha_k\, dx = 0$ when $j \neq k$ and $\int_0^\pi \alpha_j \alpha_k\, dx = \pi/2$ when $j = k$. Of course the b_k are the Fourier coefficients that we studied earlier in this chapter.

Certainly Fourier analysis has been one of the driving forces in the development of modern analysis. Questions of sets of convergence for Fourier series led to Cantor's set theory. Other convergence questions led to Dirichlet's original definition of convergent series. Riemann's theory of the integral first occurs in his classic paper on Fourier series. In turn, the tools of analysis shed much light on the fundamental questions of Fourier theory.

In more modern times, Fourier analysis was an impetus to the development of functional analysis, pseudodifferential operators, and many of the other key ideas in the subject. It continues to enjoy a symbiotic relationship with many of the newest and most incisive ideas in mathematical analysis.

One of the modern vectors in harmonic analysis is the development of wavelet theory. This is a "designer" version of harmonic analysis that allows the user to customize the building blocks. That is to say: classically, harmonic analysis taught us to build up functions from sines and cosines; wavelet theory allows us to build up functions from units that are tailored to the problem at hand. This has been shown to be a powerful tool for signal processing, signal compression, and many other contexts in which a fine and rapid analysis is desirable. In Chapter 15 we give a rapid and empirical introduction to wavelets, concentrating more on effects than on rigor. The chapter makes more than the usual demands on the reader, and certainly requires an occasional suspension of disbelief. The reward is a rich and promising theory, together with an invitatation to further reading and study.

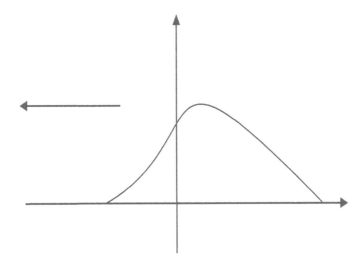

Figure 16.4: A wave of fixed shape moving to the left.

Exercises

1. Find the cosine series for the function defined on the interval $0 \le x \le 1$ by $f(x) = x^2 - x + 1/6$. This is a special instance of the Bernoulli polynomials.

Solve the following two exercises without worrying about convergence of series or differentiability of functions.

* **2.** If $y = F(x)$ is an arbitrary function, then $y = F(x + at)$ represents a wave of fixed shape that moves to the left along the x-axis with velocity a (Figure 16.4).

Similarly, if $y = G(x)$ is another arbitrary function, then $y = G(x - at)$ is a wave moving to the right, and the most general one-dimensional wave with velocity a is

$$y(x, t) = F(x + at) + G(x - at). \qquad (*)$$

(a) Show that $(*)$ satisfies the wave equation.

(b) It is easy to see that the constant a in the wave equation has the dimensions of velocity. Also, it is intuitively clear that if a stretched string is disturbed, then the waves will move in both directions away from the source of the disturbance. These considerations suggest introducing the new variables $\alpha = x + at$, $\beta = x - at$. Show that with these independent variables, the wave equation becomes

$$\frac{\partial^2 y}{\partial \alpha \partial \beta} = 0\,.$$

From this derive (*) by integration. We usually refer to (*) as *d'Alembert's solution* of the wave equation. It was also obtained, slightly later and independently, by Euler.

*   3. Consider an infinite string stretched taut on the x-axis from $-\infty$ to $+\infty$. Let the string be drawn aside into a curve $y = f(x)$ and released, and assume that its subsequent motion is described by the wave equation.

  (a) Use (*) in Exercise 2 to show that the string's displacement is given by *d'Alembert's formula*

$$y(x,t) = \frac{1}{2}[f(x + at) + f(x - at)]\,. \qquad\qquad (**)$$

  [**Hint:** Remember the initial conditions.]

  (b) Assume further that the string remains motionless at the points $x = 0$ and $x = \pi$ (such points are called *nodes*), so that $y(0,t) = y(\pi,t) = 0$, and use (**) to show that f is an odd function that is periodic with period 2π (that is, $f(-x) = -f(x)$ and $f(x+2\pi) = f(x)$).

  (c) Show that since f is odd and periodic with period 2π then f necessarily vanishes at 0 and π.

4. Solve the vibrating string problem in the text if the initial shape $y(x,0) = f(x)$ is specified by the given function. In each case, sketch the initial shape of the string on a set of axes.

  (a)

$$f(x) = \begin{cases} 2cx/\pi & \text{if } \ 0 \le x \le \pi/2 \\ 2c(\pi - x)/\pi & \text{if } \ \pi/2 \le x \le \pi \end{cases}$$

(b)

$$f(x) = \frac{1}{\pi}x(\pi - x)$$

(c)

$$f(x) = \begin{cases} x & \text{if } 0 \leq x \leq \pi/4 \\ \pi/4 & \text{if } \pi/4 < x < 3\pi/4 \\ \pi - x & \text{if } 3\pi/4 \leq x \leq \pi \end{cases}$$

* **5.** Solve the vibrating string problem in the text if the initial shape $y(x, 0) = f(x)$ is that of a single arch of the sine curve $f(x) = c\sin x$. Show that the moving string always has the same general shape, regardless of the value of c. Do the same for functions of the form $f(x) = c\sin nx$. Show in particular that there are $n - 1$ points between $x = 0$ and $x = \pi$ at which the string remains motionless; these points are called *nodes*, and these solutions are called *standing waves*. Draw sketches to illustrate the movement of the standing waves.

6. The problem of the *struck string* is that of solving the wave equation with the boundary conditions

$$y(0, t) = 0 \ , \quad y(\pi, t) = 0$$

and the initial conditions

$$\left.\frac{\partial y}{\partial t}\right|_{t=0} = g(x) \qquad \text{and} \qquad y(x, 0) = 0 \,.$$

[These initial conditions reflect the fact that the string is initially in the equilibrium position, and has an initial velocity $g(x)$ at the point x as a result of being struck.] By separating variables and proceeding formally, obtain the solution

$$y(x, t) = \sum_{j=1}^{\infty} c_j \sin jx \sin jat \,,$$

where

$$c_j = \frac{2}{\pi j a}\int_0^\pi g(x)\sin jx \, dx \,.$$

7. Solve the boundary value problem

$$
\begin{aligned}
a^2 \frac{\partial^2 w}{\partial x^2} &= \frac{\partial w}{\partial t} \\
w(x,0) &= f(x) \\
w(0,t) &= 0 \\
w(\pi,t) &= 0
\end{aligned}
$$

if the last three conditions—the boundary conditions—are changed to

$$
\begin{aligned}
w(x,0) &= f(x) \\
w(0,t) &= w_1 \\
w(\pi,t) &= w_2 .
\end{aligned}
$$

[**Hint:** Perturb the usual solution of the wave equation by a function $g(x)$.]

* **8.** Suppose that the lateral surface of the thin rod that we analyzed in the text is not insulated, but in fact radiates heat into the surrounding air. If Newton's law of cooling (that a body cools at a rate proportional to the difference of its temperature with the temperature of the surrounding air) is assumed to apply, then show that the 1-dimensional heat equation becomes

$$
a^2 \frac{\partial^2 w}{\partial x^2} = \frac{\partial w}{\partial t} + c(w - w_0)
$$

where c is a positive constant and w_0 is the temperature of the surrounding air.

* **9.** In Exercise 8, find $w(x,t)$ if the ends of the rod are kept at $0°C$, $w_0 = 0°C$, and the initial temperature distribution on the rod is $f(x)$.

10. In the solution of the heat equation, suppose that the ends of the rod are insulated instead of being kept fixed at $0°C$. What are the new boundary conditions? Find the temperature $w(x,t)$ in this case by using just common sense.

11. Solve the problem of finding $w(x,t)$ for the rod with insulated ends at $x = 0$ and $x = \pi$ (see the preceding exercise) if the initial temperature distribution is given by $w(x,0) = f(x)$.

12. The 2-dimensional heat equation is

$$a^2 \left(\frac{\partial^2 w}{\partial x^2} + \frac{\partial^2 w}{\partial y^2} \right) = \frac{\partial w}{\partial t} .$$

Use the method of separation of variables to find a steady-state solution of this equation in the infinite strip of the x-y plane bounded by the lines $x = 0$, $x = \pi$, and $y = 0$ if the following boundary conditions are satisfied:

$$w(0, y) = 0 \qquad\qquad w(\pi, y) = 0$$
$$w(x, 0) = f(x) \qquad\qquad \lim_{y \to +\infty} w(x, y) = 0 .$$

13. Derive the 3-dimensional heat equation

$$a^2 \left(\frac{\partial^2 w}{\partial x^2} + \frac{\partial^2 w}{\partial y^2} + \frac{\partial^2 w}{\partial z^2} \right) = \frac{\partial w}{\partial t}$$

by adapting the reasoning in the text to the case of a small box with edges Δx, Δy, Δz contained in a region R in x-y-z space where the temperature function $w(x, y, z, t)$ is sought. **Hint:** Consider the flow of heat through two opposite faces of the box, first perpendicular to the x-axis, then perpendicular to the y-axis, and finally perpendicular to the z-axis.

14. Solve the Dirichlet problem for the unit disc when the boundary function $f(\theta)$ is defined by

(a) $f(\theta) = \cos \theta / 2$, $\quad -\pi \le \theta \le \pi$

(b) $f(\theta) = \theta$, $\quad -\pi < \theta < \theta$

(c) $f(\theta) = \begin{cases} 0 & \text{if } -\pi \le \theta < 0 \\ \sin \theta & \text{if } 0 \le \theta \le \pi \end{cases}$

(d) $f(\theta) = \begin{cases} 0 & \text{if } -\pi \le \theta < 0 \\ 1 & \text{if } 0 \le \theta \le \pi \end{cases}$

(e) $f(\theta) = \theta^2 / 4$, $\quad -\pi \le \theta \le \pi$

15. Show that the Dirichlet problem for the disc $\{(x, y) : x^2 + y^2 \leq R^2\}$, where $f(\theta)$ is the boundary function, has the solution

$$w(r, \theta) = \frac{1}{2}a_0 + \sum_{j=1}^{\infty} \left(\frac{r}{R}\right)^j (a_j \cos j\theta + b_j \sin j\theta)$$

where a_j and b_j are the Fourier coefficients of f. Show also that the Poisson integral formula for this more general disc setting is

$$w(r, \theta) = \frac{1}{2\pi} \int_{-\pi}^{\pi} \frac{R^2 - r^2}{R^2 - 2Rr \cos(\theta - \phi) + r^2} f(\phi) \, d\phi.$$

* **16.** Let w be a harmonic function in a planar region, and let C be any circle entirely contained (along with its interior) in this region. Show that the value of w at the center of C is the average of its values on the circumference.

* **17.** If $w = F(x, y) = \mathcal{F}(r, \theta)$, with $x = r \cos \theta$ and $y = r \sin \theta$, then show that

$$\frac{\partial^2 w}{\partial x^2} + \frac{\partial^2 w}{\partial y^2} = \frac{1}{r} \left\{ \frac{\partial}{\partial r} \left(r \frac{\partial w}{\partial r} \right) + \frac{1}{r} \frac{\partial^2 w}{\partial \theta^2} \right\}$$

$$= \frac{\partial^2 w}{\partial r^2} + \frac{1}{r} \frac{\partial w}{\partial r} + \frac{1}{r^2} \frac{\partial^2 w}{\partial \theta^2}.$$

Hint: We can calculate that

$$\frac{\partial w}{\partial r} = \frac{\partial w}{\partial x} \cos \theta + \frac{\partial w}{\partial y} \sin \theta \quad \text{and} \quad \frac{\partial w}{\partial \theta} = \frac{\partial w}{\partial x} (-r \sin \theta) + \frac{\partial w}{\partial y} (r \cos \theta).$$

Similarly, compute $\dfrac{\partial}{\partial r} \left(r \dfrac{\partial w}{\partial r} \right)$ and $\dfrac{\partial^2 w}{\partial \theta^2}$.

18. Calculate the Fourier transform of $f(x) = x \cdot \chi_{[0,1]}$ where $\chi_{[0,1]}$ equals 1 at elements of the interval and 0 otherwise.

19. Calculate the Fourier transform of $g(x) = \cos x \cdot \chi_{[0,2]}$ where $\chi_{[0,2]}$ equals 1 at elements of the interval and 0 otherwise.

20. If f, g are integrable functions on $\mathbb{R}$ then define their *convolution* to be

$$h(x) = f * g(x) = \int_{\mathbb{R}} f(x - t)g(t)\, dt\,.$$

Show that

$$\widehat{h}(\xi) = \widehat{f}(\xi) \cdot \widehat{g}(\xi)\,.$$

21. Let f be a function on $\mathbb{R}$ that vanishes outside a compact set. Show that $\widehat{f}$ does *not* vanish outside any compact set.

Appendices

Glossary of Terms
from
Complex Variable Theory and Analysis

accumulation point Let $a_1, a_2, \ldots$ be points in the complex plane. A point b is an accumulation point of the a_j if the a_j get arbitrarily close to b. More formally, we require that for each $\epsilon > 0$ there exists an $N > 0$ such that when $j > N$ then $|a_j - b| < \epsilon$. §§5.2.1.

analytic continuation The procedure for enlarging the domain of a holomorphic function. §§5.2.1.

analytic continuation of a function If $(f_1, U_1), \ldots, (f_k, U_k)$ are function elements and if each (f_j, U_j) is a direct analytic continuation of (f_{j-1}, U_{j-1}), $j = 2, \ldots, k$, then we say that (f_k, U_k) is an *analytic continuation* of (f_1, U_1). §§5.2.1.

analytic continuation of a function element along a curve An *analytic continuation* of (f, U) along the curve γ is a collection of function elements (f_t, U_t), $t \in [0, 1]$, such that the f_t are compatible on the intersections of the U_ts. §§5.2.1.

annulus A set of one of the forms $\{z \in \mathbb{C} : 0 < |z| < R\}$ or $\{z \in \mathbb{C} : r < |z| < R\}$ or $\{z \in \mathbb{C} : r < |z| < \infty\}$. §§3.2.2.

argument If $z = re^{i\theta}$ is a complex number written in polar form then θ is the argument of z. §§2.1.4.

argument principle Let f be a function that is holomorphic on a domain that contains the closed disc $D(P, r)$. Assume that no zeros of f lie on $\partial D(P, r)$. Then, counting the zeros of f according to multiplicity,

$$\frac{1}{2\pi i} \oint_{\partial D(P,r)} \frac{f'(\zeta)}{f(\zeta)} \, d\zeta = \# \text{ zeros of } f \text{ inside } D(P, r).$$

Chapter 9.

argument principle for meromorphic functions Let f be a holomorphic function on a domain $U \subseteq \mathbb{C}$. Assume that $\overline{D}(P, r) \subseteq U$, and that f has neither zeros nor poles on $\partial D(P, r)$. Then

$$\frac{1}{2\pi i} \oint_{\partial D(P,r)} \frac{f'(\zeta)}{f(\zeta)} \, d\zeta = \sum_{j=1}^{p} n_j - \sum_{k=1}^{q} m_k,$$

where $n_1, n_2, \ldots, n_p$ are the multiplicities of the zeros $z_1, z_2, \ldots, z_p$ of f in $D(P, r)$ and $m_1, m_2, \ldots, m_q$ are the orders of the poles $w_1, w_2, \ldots, w_q$ of f in $D(P, r)$. §§9.1.7.

associative law If a, b, c are complex numbers then

$$(a + b) + c = a + (b + c) \qquad \text{(Associativity of Addition)}$$

and

$$(a \cdot b) \cdot c = a \cdot (b \cdot c). \qquad \text{(Associativity of Multiplication)}$$

§§1.2.3.

assumes the value β to order n A holomorphic function assumes the value β to order n at the point P if the function $f(z) - \beta$ vanishes to order n at P. §§9.1.3.

biholomorphic mapping See *conformal mapping*. §§11.1.2

Blaschke factor This is a function of the form

$$B_a(z) = \frac{z - a}{1 - \overline{a}z}$$

for some complex constant a of modulus less than one. See also *Möbius transformation.* §§5.2.3.

boundary maximum principle for harmonic functions Let $U \subseteq \mathbb{C}$ be a bounded domain. Let u be a continuous function on $\overline{U}$ that is harmonic on U. Then the maximum value of u on $\overline{U}$ (which must occur, since $\overline{U}$ is closed and bounded—see [RUD1], [KRA2]) must occur on ∂U.
Chapter 10.

boundary maximum principle for holomorphic functions Let $U \subseteq \mathbb{C}$ be a bounded domain. Let f be a continuous function on $\overline{U}$ that is holomorphic on U. Then the maximum value of $|f|$ on $\overline{U}$ (which must occur, since $\overline{U}$ is closed and bounded—see [RUD1], [KRA2]) must occur on ∂U.
Chapter 10.

boundary minimum principle for harmonic functions Let $U \subseteq \mathbb{C}$ be a bounded domain. Let u be a continuous function on $\overline{U}$ that is harmonic on U. Then the minimum value of u on $\overline{U}$ (which must occur, since $\overline{U}$ is closed and bounded—see [RUD1], [KRA2]) must occur on ∂U.
Chapter 10.

boundary uniqueness for harmonic functions Let $U \subseteq \mathbb{C}$ be a bounded domain. Let u_1 and u_2 be continuous functions on $\overline{U}$ which are harmonic on U. If $u_1 = u_2$ on ∂U then $u_1 = u_2$ on all of $\overline{U}$. §§13.2.5.

Casorati–Weierstrass theorem Let f be holomorphic on a deleted neighborhood of P and suppose that f has an essential singularity at P. Then the set of values of f is dense in the complex plane. §§6.1.6.

Cauchy estimates If f is holomorphic on a region containing the closed disc $\overline{D}(P,r)$ and if $|f| \leq M$ on $\overline{D}(P,r)$, then

$$\left| \frac{\partial^k}{\partial z^k} f(P) \right| \leq \frac{M \cdot k!}{r^k}.$$

§§5.1.2.

Cauchy integral formula Let f be holomorphic on an open set U that contains the closed disc $\overline{D}(P,r)$. Let $\gamma(t) = P + re^{it}$. Then, for each

$z \in D(P, r)$,

$$f(z) = \frac{1}{2\pi i} \oint_\gamma \frac{f(\zeta)}{\zeta - z} \, d\zeta.$$

See §§2.3.1. The formula is also true for more general curves.   §§4.2.4, 4.2.5.

Cauchy integral formula for an annulus Let f be holomorphic on an annulus $\{z \in \mathbb{C} : r < |z - P| < R\}$. Let $r < s < S < R$. Then for each $z \in D(P, S) \setminus \overline{D}(P, s)$ we have

$$f(z) = \frac{1}{2\pi i} \oint_{|\zeta - P| = S} \frac{f(\zeta)}{\zeta - P} \, d\zeta - \frac{1}{2\pi i} \oint_{|\zeta - P| = s} \frac{f(\zeta)}{\zeta - P} \, d\zeta.$$

§§6.2.5.

Cauchy integral theorem If f is holomorphic on a disc U and if $\gamma : [a, b] \to U$ is a closed curve then

$$\oint_\gamma f(z) \, dz = 0. \qquad\qquad §§2.3.2.$$

The formula is also true for more general curves.   §4.2.

Cauchy–Riemann equations If u and v are real-valued, continuously differentiable functions on the domain U then u and v are said to satisfy the Cauchy–Riemann equations on U if

$$\frac{\partial u}{\partial x} = \frac{\partial v}{\partial y} \qquad \text{and} \qquad \frac{\partial v}{\partial x} = -\frac{\partial u}{\partial y}.$$

§§3.1.2.

Cauchy–Schwarz Inequality The statement that if $z_1, \ldots z_n$ and $w_1, \ldots, w_n$ are complex numbers then

$$\left| \sum_{j=1}^n z_j w_j \right|^2 \leq \sum_{j=1}^n |z_j|^2 \sum_{j=1}^n |w_j|^2.$$

§§2.1.5.

Cayley transform This is the function

$$f(z) = \frac{i-z}{i+z}$$

that conformally maps the upper halfplane to the unit disc.   §§11.3.5.

classification of singularities in terms of Laurent series Let the holomorphic function f have an isolated singularity at P, and let

$$\sum_{j=-\infty}^{\infty} a_j(z-P)^j$$

be its Laurent expansion. Then

- If $a_j = 0$ for all $j < 0$ then f has a removable singularity at P.

- If, for some $k < 0$, $a_k \neq 0$ and $a_j = 0$ for $j < k$ then f has a pole of order k at P.

- If there are infinitely many non-zero a_j with negative index j then f has an essential singularity at P.

§§6.2.8.

clockwise The direction of traversal of a curve γ such that the region interior to the curve is always on the right.   §§4.1.1.

closed curve A curve $\gamma : [a, b] \to \mathbb{C}$ such that $\gamma(a) = \gamma(b)$.   §§4.1.2.

closed disc of radius r and center P A disc in the plane having radius r and center P and *including* the boundary of the disc.   §§1.2.2.

closed set A set E in the plane with the property that the complement of E is open.   §§1.2.2.

commutative law If a, b, c are complex numbers then

$$a + b = b + a \qquad \text{(Commutativity of Addition)}$$

and

$$a \cdot b = b \cdot a \qquad \text{(Commutativity of Multiplication)}$$

§§1.2.3.

compact A set $K \subseteq \mathbb{C}$ is compact if and only if it is both closed and bounded.   §§1.2.2.

complex derivative If f is a function on a domain U then the complex derivative of f at a point P in U is the limit

$$\lim_{z \to P} \frac{f(z) - f(P)}{z - P} .$$

§§3.1.6.

complex differentiable A function f is differentiable on a domain U if it possesses the complex derivative at each point of U.   §3.3.

complex line integral Let U be a domain, g a continuous function on U, and $\gamma : [a, b] \to U$ a curve. The complex line integral of g along γ is

$$\oint_\gamma g(z) \, dz \equiv \int_a^b g(\gamma(t)) \cdot \frac{d\gamma}{dt}(t) \, dt .$$

§4.1.

complex numbers Any number of the form $x + iy$ with x and y real. §§1.1.2.

conformal A function f on a domain U is conformal if it preserves angles and dilates equally in all directions. A holomorphic function is conformal, and conversely.   §§3.3.1.

conformal mapping Let U, V be domains in $\mathbb{C}$. A function $f : U \to V$ that is holomorphic, one-to-one, and onto is called a conformal mapping or conformal map.   §§11.1.

conformal self-map Let $U \subseteq \mathbb{C}$ be a domain. A function $f : U \to U$ that is holomorphic, one-to-one, and onto is called a conformal (or bi-holomorphic) self-map of U. §11.1, 11.2.

conjugate If $z = x + iy$ is a complex number then $\overline{z} = x - iy$ is its conjugate. §§1.1.3.

connected A set S in the plane is connected if there do not exist disjoint and non-empty open sets U and V such that $S = (S \cap U) \cup (S \cap V)$. §§1.2.2.

continuous A function f with domain S is continuous at a point P in S if the limit of $f(x)$ as x approaches P is $f(P)$. An equivalent definition, coming from topology, is that f is continuous provided that whenever V is an open set in the range of f then $f^{-1}(V)$ is open in the domain of f. §§3.1.1.

continuously differentiable A function f with domain S is continuously differentiable if the first derivative(s) of f exist at every point of S and if each of those first derivative functions is continuous on S. §§3.1.1.

continuously differentiable, k times A function f with domain S such that all derivatives of f up to and including order k exist and each of those derivative functions is continuous on S. §§3.1.1.

convergence of a Laurent series The Laurent series

$$\sum_{j=-\infty}^{\infty} a_j(z - P)^j$$

is said to converge if each of the power series

$$\sum_{j=-\infty}^{0} a_j(z - P)^j \quad \text{and} \quad \sum_{1}^{\infty} a_j(z - P)^j$$

converges. §6.2.

convergence of a power series The power series

$$\sum_{j=0}^{\infty} a_j(z - P)^j$$

is said to converge at z if the partial sums $S_N(z)$ converge as a sequence
of numbers.   §5.1.

converges uniformly See *uniform convergence.*   §§5.1.5

countable set A set S is countable if there is a one-to-one, onto function
$f : S \to \mathbb{N}$.   §§11.5.3

countably infinite set See *countable set.*   §§11.5.3

counterclockwise The direction of traversal of a curve γ such that the
region interior to the curve is always on the left. §§2.1.7.

counting function This is a function from classical number theory that
aids in counting the prime numbers. §§12.3.1.

curve A continuous function $\gamma : [a, b] \to \mathbb{C}$.   §§4.1.1.

deformability Let U be a domain. Let $\gamma : [a, b] \to U$ and $\mu : [a, b] \to U$
be curves in U. We say that γ is deformable to μ in U if there is a continuous function $H(s, t)$, $0 \leq s \leq 1$ such that $H(0, t) = \gamma(t)$, $H(1, t) = \mu(t)$,
and $H(s, t) \in U$ for all (s, t).   §§4.2.3.

deleted neighborhood Let $P \in \mathbb{C}$. A set of the form $D(P, r) \setminus \{P\}$ is
called a deleted neighborhood of P.   §§6.1.2.

derivative with respect to z If f is a function on a domain U then
the derivative of f with respect to z on U is

$$\frac{\partial f}{\partial z} = \frac{1}{2}\left(\frac{\partial}{\partial x} - i\frac{\partial}{\partial y}\right) f.$$

§§3.1.3.

derivative with respect to $\bar{z}$ If f is a function on a domain U then the derivative of f with respect to $\bar{z}$ on U is

$$\frac{\partial f}{\partial \bar{z}} = \frac{1}{2}\left(\frac{\partial}{\partial x} + i\frac{\partial}{\partial y}\right) f.$$

§§3.1.3.

differentiable See *complex differentiable*. §§3.1.7.

Dirichlet problem on the disc Given a continuous function f on $\partial D(0,1)$, find a continuous function u on $\overline{D}(0,1)$ whose restriction to $\partial D(0,1)$ equals f. §12.2.

Dirichlet problem on a general domain Let $U \subseteq \mathbb{C}$ be a domain. Let f be a continuous function on ∂U. Find a continuous function u on $\overline{U}$ such that u agrees with f on ∂U. §12.2, 13.3, 16.1.

disc of convergence A power series

$$\sum_{j=0}^{\infty} a_j(z - P)^j$$

converges on a disc $D(P,r)$, where

$$r = \frac{1}{\limsup_{j\to\infty} |a_j|^{1/j}}.$$

The disc $D(P,r)$ is the disc of convergence of the power series. §§5.1.6.

discrete set A set $S \subset \mathbb{C}$ is discrete if for each $s \in S$ there is an $\delta > 0$ such that $D(s,\delta) \cap S = \{s\}$. See also *isolated point*. §§5.2.2.

distributive law If a, b, c are complex numbers then the distributive laws are

$$a \cdot (b + c) = ab + ac$$

and

$$(b + c) \cdot a = ba + ca.$$

§§1.2.3.

domain A set U in the plane that is both open and connected. §§1.2.2.

domain of a function The domain of a function f is the set of numbers or points to which f can be applied.   §§6.1.2

entire function A holomorphic function whose domain consists of the complex plane $\mathbb{C}$.   §§5.1.3.

essential singularity If the point P is a singularity of the holomorphic function f, and if P is neither a removable singularity nor a pole, then P is called an essential singularity. §§4.1.4,   §§6.1.4.

Euclidean algorithm The algorithm for long division in the theory of arithmetic.   §§1.2.4.

exponential, complex The function e^z.   §2.1.

extended plane The complex plane with the point at infinity adjoined. See *stereographic projection*.   §§7.2.9.

field A number system that is closed under addition, multiplication, and division by non-zero numbers and in which these operations are commutative.   §§1.2.3.

formula for the derivative Let $U \subseteq \mathbb{C}$ be an open set and let f be holomorphic on U. Then f is infinitely differentiable on U. Moreover, if $\overline{D}(P, r) \subseteq U$ and $z \in D(P, r)$ then

$$\left(\frac{\partial}{\partial z}\right)^k f(z) = \frac{k!}{2\pi i} \oint_{|\zeta - P| = r} \frac{f(\zeta)}{(\zeta - z)^{k+1}} \, d\zeta, \quad k = 0, 1, 2, \ldots.$$

§§5.1.1.

Fundamental Theorem of Algebra The statement that every non-constant polynomial has a root. §§1.2.4.

Fundamental Theorem of Calculus along Curves Let $U \subset \mathbb{C}$ be a domain and $\gamma = (\gamma_1, \gamma_2) : [a, b] \to U$ a C^1 curve. If $f \in C^1(U)$ then

$$f(\gamma(b)) - f(\gamma(b)) = \int_a^b \left(\frac{\partial f}{\partial x}(\gamma(t)) \cdot \frac{d\gamma_1}{dt} + \frac{\partial f}{\partial y}(\gamma(t)) \cdot \frac{d\gamma_2}{dt} \right) dt.$$

§§4.1.5.

generalized circles and lines In the extended plane $\widehat{\mathbb{C}} = \mathbb{C} \cup \{\infty\}$, a generalized line (generalized circle) is an ordinary line union the point at infinity. Topologically, an extended line is a circle. §§11.3.6.

harmonic A function u on a domain U is said to be harmonic of $\triangle u = 0$ on U, that is, if u satisfies the Laplace equation. Chapter 3.

harmonic conjugate If u is a real-valued harmonic function on a domain u then a real-valued harmonic function v on U is said to be conjugate to u if $h = u + iv$ is holomorphic. §§3.2.2.

holomorphic A continuously differentiable function on a domain U is holomorphic if it satisfies the Cauchy–Riemann equations on U or (equivalently) if $\partial f / \partial \overline{z} = 0$ on U. Chapter 3.

holomorphic function on a Riemann surface A function F is holomorphic on the Riemann surface $\mathcal{R}$ if $F \circ \pi^{-1} : \pi(U) \to \mathbb{C}$ is holomorphic for each open set U in $\mathbb{R}$ with π one-to-one on U. §§11.5.3.

homeomorphic Two open sets U and V in $\mathbb{C}$ are *homeomorphic* if there is a one-to-one, onto, continuous function $f : U \to V$ with $f^{-1} : V \to U$ also continuous. §§11.4.1.

homeomorphism A *homeomorphism* of two sets $A, B \subseteq \mathbb{C}$ is a one-to-one, onto continuous mapping $F : A \to B$ which has a continuous inverse.   §§11.4.1.

homotopic See *deformability, homotopy.*   §§4.2.2.

Hurwitz's theorem Suppose that $U \subseteq \mathbb{C}$ is a domain and that $\{f_j\}$ is a sequence of nowhere-vanishing holomorphic functions on U. If the sequence $\{f_j\}$ converges uniformly on compact subsets of U to a (necessarily holomorphic) limit function f_0 then either f_0 is nowhere-vanishing or $f_0 \equiv 0$.   §§9.3.4.

image of a function The set of values taken by the function. §§4.1.1.

imaginary part If $z = x + iy$ is a complex number then its imaginary part is y.   §§1.1.2.

imaginary part of a function f If $f = u + iv$ is a complex-valued function, with u and v real-valued functions, then the function v is its imaginary part.   §§3.1.2.

index Let U be a domain and $\gamma : [0,1] \to U$ a piecewise C^1 curve in U. Let $P \in U$ be a point that does not lie on γ. Then the index of γ with respect to P is defined to be

$$\mathrm{Ind}_\gamma(P) \equiv \frac{1}{2\pi i} \oint_\gamma \frac{1}{\zeta - P} \, d\zeta.$$

The index is always an integer.   §§8.1.5.

isolated point A point P of a set $S \subseteq \mathbb{C}$ is said to be isolated if there is an $\delta > 0$ such that $D(P, \delta) \cap S = \{P\}$.   §§7.2.2.

isolated singularity See *singularity.*   Chapter 6.

isolated singular point See *singularity.*   §6.1.

k times continuously differentiable A function f with domain S such that all derivatives of f up to and including order k exist and each of those derivatives is continuous on S. §§3.1.1.

Laplace equation The partial differential equation

$$\triangle u = 0.$$ §§7.1.1

§§13.1.1.

Laplace operator or Laplacian This is the partial differential operator

$$\triangle = \frac{\partial^2}{\partial x^2} + \frac{\partial^2}{\partial x^2}.$$

§§7 3.2.1.

Laurent series A series of the form

$$\sum_{j=-\infty}^{\infty} a_j(z - P)^j.$$

See also *power series*. §§6.2.1.

Laurent series expansion about ∞ Fix a positive number R. Let f be holomorphic on a set of the form $\{z \in \mathbb{C} : |z| > R\}$. Define $G(z) = f(1/z)$ for $|z| < 1/R$. If the Laurent series expansion of G about 0 is

$$\sum_{j=-\infty}^{\infty} a_j z^j$$

then the Laurent series expansion of f about ∞ is

$$\sum_{j=-\infty}^{\infty} a_j z^{-j}.$$

§§7.2.7.

limit of the function f at the point P Let f be a function on a domain U. The complex number ℓ is the limit of the f at P if for each $\epsilon > 0$ there is a $\delta > 0$ such that whenever $z \in U$ and $0 < |z - P| < \delta$

then $|f(z) - P| < \epsilon$.   §§3.1.6.

linear fractional transformation A function of the form

$$z \mapsto \frac{az + b}{cz + d},$$

for a, b, c, d complex constants with $ac - bd \neq 0$.   §§11.3.

Liouville's theorem If f is an entire function that is bounded then f is constant.   §§5.1.3.

maximum principle for harmonic functions If u is a harmonic function on a domain U and if P in U is a local maximum for u then u is identically constant.   §§13.2.1.

maximum principle for holomorphic functions If f is a holomorphic function on a domain U and if P in U is a local maximum for $|f|$ then f is identically constant.   §10.1.

mean value property for harmonic functions Let u be harmonic on an open set containing the closed disc $\overline{D}(P, r)$. Then

$$u(P) = \frac{1}{2\pi} \int_0^{2\pi} u(P + re^{i\theta}) \, d\theta.$$

This identity also holds for holomorphic functions.   §§13.2.4.

meromorphic at ∞ Fix a positive number R. Let f be holomorphic on a set of the form $\{z \in \mathbb{C} : |z| > R\}$. Define $G(z) = f(1/z)$ for $|z| < 1/R$. We say that f is meromorphic at ∞ provided that G is meromorphic in the usual sense on $\{z \in \mathbb{C} : |z| < 1/R\}$.   §§7.2.8.

meromorphic function Let U be a domain and $\{P_j\}$ a discrete set in U. If f is holomorphic on $U \setminus \{P_j\}$ and f has a pole at each of the $\{P_j\}$ then f is said to be meromorphic on U.   Chapter 7.

minimum principle for harmonic functions If u is a harmonic function on a domain U and if P in U is a local minimum for u then u is identically constant. §§13.2.2.

minimum principle for holomorphic functions If f is a holomorphic function on a domain U, if f does not vanish on U, and if P in U is a local minimum for $|f|$ then f is identically constant. §§10.1.3.

Möbius transformation This is a function of the form

$$\phi_a(z) = \frac{z - a}{1 - \bar{a}z}$$

for a fixed complex constant a with modulus less than 1. Such a function ϕ_a is a conformal self-map of the unit disc. §§11.2.2.

modulus If $z = x + iy$ is a complex number then $|z| = \sqrt{x^2 + y^2}$ is its modulus. §§1.2.1.

monogenic See *holomorphic*. §§3.1.7.

Morera's theorem Let f be a continuous function on a connected open set $U \subseteq \mathbb{C}$. If

$$\oint_\gamma f(z)\,dz = 0$$

for every simple closed curve γ in U then f is holomorphic on U. The result is true if it is only assumed that the integral is zero when γ is a rectangle, or when γ is a triangle. §§4.2.1.

multiple root Let f be either a polynomial or a holomorphic function on an open set U. Let k be a positive integer. If $P \in U$ and $f(P) = 0, f'(P) = 0, \ldots, f^{(k-1)}(P) = 0$ then f is said to have a multiple root at P. The root is said to be of order k. §§9.1.7.

multiple singularities Let $U \subseteq \mathbb{C}$ be a domain and $P_1, P_2, \ldots$ be a discrete set in U. If f is holomorphic on $U \setminus \{P_j\}$ and has a singularity at each P_j then f is said to have multiple singularities in U. §§8.1.1.

multiplicity of a zero or root The number k in the definition of *multiple root*.   §§5.1.4.

one-to-one A function $f : S \to T$ is said to be one-to-one if whenever $s_1 \neq s_2$ then $f(s_1) \neq f(s_2)$.   §§4.1.2.

open disc of radius r and center P A disc $D(P, r)$ in the plane having radius r and center P and *not* including the circle which is the boundary of the disc.   §§1.2.2.

open mapping A function $f : S \to T$ is said to be open if whenever $U \subseteq S$ is open then $f(U) \subseteq T$ is open.   §§9.2.1.

open mapping theorem If $f : U \to \mathbb{C}$ is a holomorphic function on a domain U, then $f(U)$ will also be open.   §§9.2.1.

open set A set U in the plane with the property that each point $P \in U$ has a disc $D(P, r)$ such that $D(P, r) \subseteq U$.   §§1.2.2.

order of a pole See *pole*.   §§6.1.4.

order of a root See *multiplicity of a root*.   §§5.1.4.

partial fractions A method for decomposing a rational function into a sum of simpler rational components. Useful in integration theory, as well as in various algebraic contexts. See [THO] for details.   §§15.1.2.

partial sums of a power series If

$$\sum_{j=0}^{\infty} a_j (z - P)^j$$

is a power series then its partial sums are the expressions

$$S_N(z) \equiv \sum_{j=0}^{N} a_j (z - P)^j$$

for $N = 0, 1, 2, \ldots$. §§5.1.6.

path See curve. §§1.2.2.

path-connected Let $E \subseteq \mathbb{C}$ be a set. If, for any two points A and B in E there is a curve $\gamma : [0, 1] \to E$ such that $\gamma(0) = A$ and $\gamma(1) = B$ then we say that E is path-connected. §§1.2.2.

piecewise C^k A curve $\gamma : [a, b] \to \mathbb{C}$ is said to be piecewise C^k if

$$[a, b] = [a_0, a_1] \cup [a_1, a_2] \cup \cdots \cup [a_{m-1}, a_m]$$

with $a = a_0 < a_1 < \cdots a_m = b$ and the restriction $\gamma\big|_{[a_{j-1}, a_j]}$ is C^k for $1 \leq j \leq m$. §§3.1.1.

point at ∞ A point which is adjoined to the complex plane to make it topologically a sphere. §§13.3.1.

Poisson integral formula Let $u : U \to \mathbb{R}$ be a harmonic function on a neighborhood of $\overline{D}(0, 1)$. Then, for any point $a \in D(0, 1)$,

$$u(a) = \frac{1}{2\pi} \int_0^{2\pi} u(e^{i\psi}) \cdot \frac{1 - |a|^2}{|a - e^{i\psi}|^2} \, d\psi.$$

§§13.3.

Poisson kernel for the unit disc This is the function

$$\frac{1}{2\pi} \frac{1 - |a|^2}{|a - e^{i\psi}|^2}$$

that occurs in the Poisson integral formula. §§13.3.2.

polar form of a complex number A complex number z written in the form $z = re^{i\theta}$ with $r \geq 0$ and $\theta \in \mathbb{R}$. The number r is the modulus of z and θ is its argument. §§2.1.2.

polar representation of a complex number See *polar form*. §§2.1.2.

pole Let P be an isolated singularity of the holomorphic function f. If P is not a removable singularity for f but there exists a $k > 0$ such that $(z - P)^k \cdot f$ is a removable singularity, then P is called a pole of f. The least integer k for which this condition holds is called the order of the pole.   §§6.1.4.

polynomial A polynomial is a function $p(z)$ (resp. $p(x)$) of the form

$$p(z) = a_0 + a_1 z + \cdots a_{k-1} z^{k-1} + a_k z^k,$$

(resp. $p(x) = a_0 + a_1 x + \cdots a_{k-1} x^{k-1} + a_k x^k$) where $a_0, \ldots, a_k$ are complex constants.   §§1.1.2.

power series A series of the form

$$\sum_{j=0}^{\infty} a_j (z - P)^j.$$

More generally, the series can have any limits on the indices:

$$\sum_{j=m}^{\infty} a_j (z - P)^j \qquad \text{or} \qquad \sum_{j=m}^{n} a_j (z - P)^j .$$

§§5.1.6.

pre-vertices The inverse images of the corners of the polygon under study with the Schwarz–Christoffel mapping.   §§12.4.1.

principal branch Usually that branch of a holomorphic function that focuses on values of the argument $0 \leq \theta < 2\pi$. The precise definition of the phrase "principal branch" depends on the particular function being studied.   §§2.1.4.

principle of persistence of functional relations If two holomorphic functions defined in a domain containing the real axis agree for real values of the argument then they agree at all points.   §§5.2.3.

principal part Let f have a pole of order k at P. The negative power part

$$\sum_{j=-k}^{-1} a_j(z-P)^j$$

of the Laurent series of f about P is called the principal part of f at P.
§§7.1.1.

rational function A *rational* function is a function that is quotient of polynomials.
§§7.2.9.

real number system Those numbers consisting of either terminating or non-terminating decimal expansions.
§§1.1.1.

real part If $z = x + iy$ is a complex number then its real part is the number x.
§§1.1.2.

real part of a function f If $f = u+iv$, with u, v real-valued functions, is a complex-valued function then u is its real part.
§§13.1.4.

region See *domain*.
§§1.2.2.

regular See *holomorphic*.
§§3.1.7

removable singularity Let P be an isolated singularity of the holomorphic function f. If f can be defined at P so as to be holomorphic in a neighborhood of P then the point P is called a removable singularity for f.
§§6.1.4.

residue If f has Laurent series

$$\sum_{j=-\infty}^{\infty} a_j(z-P)^j$$

about P, then the number a_{-1} is called the residue of f at P. We denote the residue by $\text{Res}_f(P)$.
§8.1.

residue, formula for Let f have a pole of order k at P. Then the residue of f at P is given by

$$\text{Res}_f(P) = \frac{1}{(k-1)!} \left(\frac{\partial}{\partial z}\right)^{k-1} \left((z-P)^k f(z)\right)\Big|_{z=P}.$$

§§8.1.

residue theorem Let U be a domain and let the holomorphic function f have isolated singularities at $P_1, P_2, \ldots, P_m \in U$. Let $\text{Res}_f(P_j)$ be the residue of f at P_j. Also let $\gamma : [0,1] \to U \setminus \{P_1, P_2, \ldots, P_m\}$ be a piecewise C^1 curve. Let $\text{Ind}_\gamma(P_j)$ be the winding number of γ about P_j. Then

$$\oint_\gamma f(z)\, dz = 2\pi i \sum_{j=1}^m \text{Res}_f(P_j) \cdot \text{Ind}_\gamma(P_j).$$

§§8.1.3.

Riemann mapping theorem Let $U \subseteq \mathbb{C}$ be a simply connected domain, and assume that $U \neq \mathbb{C}$. Then there is a conformal mapping $\varphi : U \to D(0,1)$.   §11.4.

Riemann removable singularities theorem If P is an isolated singularity of the holomorphic function f and if f is bounded in a deleted neighborhood of P then f has a removable singularity at P.   §§6.1.5.

Riemann sphere See *extended plane*.   §§11.3.3.

Riemann surface The idea of a Riemann surface is that one can visualize geometrically the behavior of function elements and their analytic continuation. A global analytic function is the set of all function elements obtained by analytic continuation along curves (from a base point $P \in \mathbb{C}$) of a function element (f, U) at P. Such a set, which amounts to a collection of convergent power series at different points of the plane $\mathbb{C}$, can be given the structure of a surface, in the intuitive sense of that word.   §§11.5.3.

right-turn angle Angles in the Schwarz–Christoffel mapping construction. 12.4.1.

root of a function or polynomial A value in the domain at which the
function or polynomial vanishes. See also *zero*. §§1.1.2.

rotation A function $z \mapsto e^{i\alpha}z$ for some fixed real number α. We sometimes say that the function represents "rotation through an angle of α
radians." §§3.2.2.

Rouché's theorem Let f, g be holomorphic functions on a domain $U \subseteq$
$\mathbb{C}$. Suppose that $\overline{D}(P, r) \subseteq U$ and that, for each $\zeta \in \partial D(P, r)$,

$$|f(\zeta) - g(\zeta)| < |f(\zeta)| + |g(\zeta)|. \qquad (*)$$

Then the number of zeros of f inside $D(P, r)$ equals the number of zeros
of g inside $D(P, r)$. The hypothesis $(*)$ is sometimes replaced in practice
with

$$|f(\zeta) - g(\zeta)| < |g(\zeta)|$$

for $\zeta \in \partial D(P, r)$. §§9.3.1.

Schwarz–Christoffel mapping A conformal mapping from the upper
halfplane to a polygon. §§12.4.1

Schwarz–Christoffel parameter problem The problem of determin-
ing the pre-vertices of a Schwarz–Christoffel mapping. §§12.4.1.

Schwarz lemma Let f be holomorphic on the unit disc. Assume that

1. $|f(z)| \leq 1$ for all z.

2. $f(0) = 0$.

Then $|f(z)| \leq |z|$ and $|f'(0)| \leq 1$.
    If either $|f(z)| = |z|$ for some $z \neq 0$ or if $|f'(0)| = 1$ then f is a rota-
tion: $f(z) \equiv \alpha z$ for some complex constant α of unit modulus. §10.2.

Schwarz–Pick lemma Let f be holomorphic on the unit disc. Assume that

1. $\quad |f(z)| \le 1$ for all z.

2. $\quad f(a) = b$ for some $a, b \in D(0, 1)$.

Then
$$|f'(a)| \le \frac{1 - |b|^2}{1 - |a|^2}.$$
Moreover, if $f(a_1) = b_1$ and $f(a_2) = b_2$ then
$$\left| \frac{b_2 - b_1}{1 - \overline{b_1} b_2} \right| \le \left| \frac{a_2 - a_1}{1 - \overline{a_1} a_2} \right|.$$
There is a "uniqueness" result in the Schwarz–Pick lemma. If either
$$|f'(a)| = \frac{1 - |b|^2}{1 - |a|^2} \quad \text{or} \quad \left| \frac{b_2 - b_1}{1 - \overline{b_1} b_2} \right| = \left| \frac{a_2 - a_1}{1 - \overline{a_1} a_2} \right|$$
then the function f is a conformal self-mapping (one-to-one, onto holomorphic function) of $D(0, 1)$ to itself.   §§10.2.2.

simple closed curve A curve $\gamma : [a, b] \to \mathbb{C}$ such that $\gamma(a) = \gamma(b)$ but the curve crosses itself nowhere else.   §§4.2.5.

simple root Let f be either a polynomial or a holomorphic function on an open set U. If $f(P) = 0$ but $f'(P) \ne 0$ then f is said to have a simple root at P. See also *multiple root*.   §§5.1.4.

simply connected A domain U in the plane is simply connected if one of the following three equivalent conditions holds: it has no holes, or if its complement has only one connected component, or if each closed curve in U is homotopic to zero.   §§3.2.2.

singularity Let f be a holomorphic function on $D(P, r) \setminus \{P\}$ (that is, on the disc minus its center). Then the point P is said to be a singularity of f.   Chapter 6.

singularity at ∞ Fix a positive number R. Let f be holomorphic on the set $\{z \in \mathbb{C} : |z| > R\}$. Define $G(z) = f(1/z)$ for $|z| < 1/R$. Then

- If G has a removable singularity at 0 then we say that f has a removable singularity at ∞.

- If G has a pole at 0 then we say that f has a pole at ∞.

- If G has an essential singularity at 0 then we say that f has an essential singularity at ∞.

$$\S\S 7.2.6.$$

smooth curve A curve $\gamma : [a, b] \to \mathbb{C}$ is smooth if γ is a C^k function (where k suits the problem at hand, and may be ∞) and γ' never vanishes. $\S\S 12.2.1.$

solution of the Dirichlet problem on the disc Let f be a continuous function on $\partial D(0, 1)$. Define

$$u(z) = \begin{cases} \dfrac{1}{2\pi} \displaystyle\int_0^{2\pi} f(e^{i\psi}) \cdot \dfrac{1 - |z|^2}{|z - e^{i\psi}|^2} \, d\psi & \text{if } z \in D(0, 1) \\ f(z) & \text{if } z \in \partial D(0, 1). \end{cases}$$

Then u is continuous on $\overline{D}(0, 1)$ and harmonic on $D(0, 1)$. $\S\S 13.3.4.$

stereographic projection A geometric method for mapping the plane to a sphere. $\S\S 11.3.3.$

sub-mean value property Let $f : U \to \mathbb{R}$ be continuous. Then f satisfies the sub-mean value property if, for each $\overline{D}(P, r) \subseteq U$,

$$f(P) \leq \frac{1}{2\pi} \int_0^{2\pi} f(P + re^{i\theta}) d\theta \,.$$

$$\S\S 13.2.5.$$

topology A mathematical structure specifying open and closed sets and a notion of convergence. $\S\S 1.2.2.$

triangle inequality The statement that if z, w are complex numbers then

$$|z + w| \leq |z| + |w| \,.$$

§§1.2.1.

uniform convergence for a sequence Let f_j be a sequence of functions on a set S. The f_j are said to converge uniformly to a function g on S if for each $\epsilon > 0$ there is a $N > 0$ such that if $j > N$ then $|f_j(s) - g(s)| < \epsilon$ for all $s \in S$. In other words, $f_j(s)$ converges to $g(s)$ at the same rate at each point of S. §§5.1.5.

uniform convergence for a series The series

$$\sum_{j=1}^{\infty} f_j(z)$$

on a set S is said to converge uniformly to a limit function $F(z)$ if its sequence of partial sums converges uniformly to F. Equivalently, the series converges uniformly to F if for each $\epsilon > 0$ there is a number $N > 0$ such that if $J > N$ then

$$\left| \sum_{j=1}^{J} f_j(z) - F(z) \right| < \epsilon$$

for all $z \in S$. §§5.1.5.

uniform convergence on compact subsets for a sequence Let f_j be a sequence of functions on a domain U. The f_j are said to converge uniformly on compact subsets of U to a function g on U if, for each compact $K \subseteq U$ and for each $\epsilon > 0$, there is a $N > 0$ such that if $j > N$ then $|f_j(k) - g(k)| < \epsilon$ for all $k \in K$. In other words, $f_j(k)$ converges to $g(k)$ at the same rate at each point of K. §§5.1.5.

uniform convergence on compact subsets for a series The series

$$\sum_{j=1}^{\infty} f_j(z)$$

on a domain U is said to be uniformly convergent on compact sets to a limit function $F(z)$ if, for each $\epsilon > 0$ and each compact $K \subseteq U$, there is an $N > 0$ such that if $J > N$ then

$$\left| \sum_{j=1}^{N} f(z) - F(z) \right| < \epsilon$$

for every $z \in K$. §§5.1.5.

uniqueness of analytic continuation Let f and g be holomorphic functions on a domain U. If there is a disc $D(P, r) \subseteq U$ such that f and g agree on $D(P, r)$ then f and g agree on all of U. More generally, if f and g agree on a set with an accumulation point in U then they agree at all points of U. §§5.2.3.

winding number See *index*. §§8.1.5.

zero If f is a polynomial or a holomorphic function on an open set U then $P \in U$ is a zero of f if $f(P) = 0$. See *root of a function or polynomial*. §5.2.

zero set If f is a polynomial or a holomorphic function on an open set U then the zero set of f is $\{z \in U : f(z) = 0\}$. §§5.2.1.

List of Notation

Notation	Meaning	Subsection
$\mathbb{R}$	real number system	1.1.1
$\mathbb{R}^2$	Cartesian plane	1.1.1
$\mathbb{C}$	complex number system	1.1.2
z, w, ζ	complex numbers	1.1.2
$z = x + iy$	complex numbers	1.1.2
$w = u + iv$	complex numbers	1.1.2
$\zeta = \xi + i\eta$	complex numbers	1.1.2
$\operatorname{Re} z$	real part of z	1.1.2
$\operatorname{Im} z$	imaginary part of z	1.1.2
$\overline{z}$	conjugate of z	1.1.2
$|z|$	modulus of z	1.2.1
$D(P, r)$	open disc	1.2.2
$\overline{D}(P, r)$	closed disc	1.2.2
D	open unit disc	1.2.2
$\overline{D}$	closed unit disc	1.2.2
$A \setminus B$ the complement of B in A		1.2.2
e^z	complex exponential	2.1
$!$	factorial	2.1
$\cos z$	$\frac{e^{iz} + e^{-iz}}{2}$	2.1.1
$\sin z$	$\frac{e^{iz} - e^{-iz}}{2i}$	2.1.1
$\arg z$	argument of z	2.1.4
C^k	k times continuously differentiable	3.1.1
$f = u + iv$	real and imaginary parts of f	3.1.2
$\operatorname{Re} f$	real part of the function f	3.1.2
$\operatorname{Im} f$	imaginary part of the function f	3.1.2
$\partial f / \partial z$	derivative with respect to z	3.1.3
$\partial f / \partial \overline{z}$	derivative with respect to $\overline{z}$	3.1.3
$\lim_{z \to P} f(z)$	limit of f at the point P	3.1.6
df / dz	complex derivative	3.1.6
$f'(z)$	complex derivative	3.1.6

Notation	Meaning	Subsection
$\triangle$	the Laplace operator	3.2.1
γ	a curve	4.1.1
$\gamma\big|_{[c,d]}$	restriction of γ to $[c,d]$	4.1.1
$\oint_\gamma g(z)\,dz$	complex line integral of g along γ	4.1.6
$S_N(z)$	partial sum of a power series	5.1.6
$A \setminus B$	the complement of B in A	1.2.2
$\mathrm{Res}_f(P)$	residue of f at P	8.1.2
$\mathrm{Ind}_\gamma(P)$	index of γ with respect to P	8.1.5
$\widehat{\mathbb{C}}$	the extended complex plane	7.2
$\mathbb{C} \cup \{\infty\}$	the extended complex plane	7.2
$L \cup \{\infty\}$	generalized circle	11.3.6
$B_a(z)$	Blaschke factor	5.2.3
$B(z)$	Blaschke product	5.2.3
$H : [0,1] \times [0,1] \to W$	a homotopy in W	4.2.2
θ_j	right-turn angle	12.4.1
$\widehat{f}(n)$	Fourier coefficient of f	14.1.2
$Sf(t)$	Fourier series of f	14.1.2
$S_N f(t)$	partial sum of Fourier series of f	14.1.2
$\widehat{f}(\xi)$	Fourier transform of f	14.2.1
$\overset{\vee}{g}$	inverse Fourier transform of g	14.2.1
$F(s)$	Laplace transform of f	15.1.1
$\mathcal{L}(f)$	Laplace transform of f	15.1.1
$A(z)$	z-transform of $\{a_j\}$	15.2

Table of Laplace Transforms

Function	Laplace Transform	Domain of Convergence
e^{at}	$\frac{1}{s-a}$	$\{s : \operatorname{Re} s > \operatorname{Re} a\}$
1	$\frac{1}{s}$	$\{s : \operatorname{Re} s > 0\}$
$\cos \omega t$	$\frac{s}{s^2+\omega^2}$	w real, $\{s : \operatorname{Re} s > 0\}$
$\sin \omega t$	$\frac{\omega}{s^2+\omega^2}$	w real, $\{s : \operatorname{Re} s > 0\}$
$\cosh \omega t$	$\frac{s}{s^2-\omega^2}$	w real, $\{s : \operatorname{Re} s > |\omega|\}$
$\sinh \omega t$	$\frac{\omega}{s^2-\omega^2}$	ω real, $\{s : \operatorname{Re} s > |\omega|\}$
$e^{-\lambda t} \cos \omega t$	$\frac{s+\lambda}{(s+\lambda)^2+\omega^2}$	ω, λ real, $\{s : \operatorname{Re} s > -\lambda\}$
$e^{-\lambda t} \sin \omega t$	$\frac{\omega}{(s+\lambda)^2+\omega^2}$	ω, λ real, $\{s : \operatorname{Re} s > -\lambda\}$
$t^n e^{at}$	$\frac{n!}{(s-a)^{n+1}}$	$\{s : \operatorname{Re} s > \operatorname{Re} a\}$
t^n	$\frac{n!}{s^{n+1}}$	$\{s : \operatorname{Re} s > 0\}$
$t \cos \omega t$	$\frac{s^2-\omega^2}{(s^2+\omega^2)^2}$	ω real, $\{s : \operatorname{Re} s > 0\}$
$t \sin \omega t$	$\frac{2\omega s}{(s^2+\omega^2)^2}$	ω real, $\{s : \operatorname{Re} s > 0\}$
$f'(t)$	$s\mathcal{L}f(s) - f(0)$	$\{s : \operatorname{Re} s > 0\}$
$f''(t)$	$s^2\mathcal{L}f(s) - sf(0) - f'(0)$	$\{s : \operatorname{Re} s > 0\}$
$tf(t)$	$-(\mathcal{L}f)'(s)$	$\{s : \operatorname{Re} s > 0\}$
$e^{at}f(t)$	$\mathcal{L}f(s - a)$	$\{s : \operatorname{Re} s > 0\}$

A Guide to the Literature

Complex analysis is an old subject, and the associated literature is large. Here we give the reader a representative sampling of some of the resources that are available. Of course no list of this kind can be complete.

Traditional Texts

- L. V. Ahlfors, *Conformal Invariants*, McGraw-Hill, 1973.

- C. Carathéodory, *Theory of Functions of a Complex Variable*, Chelsea, New York, 1954.

- H. P. Cartan, *Elementary Theory of Analytic Functions of One and Several Complex Variables*, Addison-Wesley, Reading, 1963.

- E. T. Copson, *An Introduction to the Theory of Functions of One Complex Variable*, The Clarendon Press, Oxford, 1972.

- R. Courant, *The Theory of Functions of a Complex Variable*, New York University, New York, 1949.

- H. Dym and H. McKean, *Fourier Series and Integrals*, Academic Press, New York, 1972.

- P. Franklin, *Functions of Complex Variables*, Prentice-Hall, Englewood Cliffs, 1959.

- W. H. J. Fuchs, *Topics in the Theory of Functions of One Complex Variable*, Van Nostrand, Princeton, 1967.

- B. A. Fuks, *Functions of a Complex Variable and Some of Their Applications*, Addison-Wesley, Reading, 1961.

- G. M. Goluzin, *Geometric Theory of Functions of a Complex Variable*, American Mathematical Society, Providence, 1969.

- K. Knopp, *Theory of Functions*, Dover, New York, 1945-1947.

- Z. Nehari, *Introduction to Complex Analysis*, Allyn & Bacon, Boston, 1961.

337

- R. Nevanlinna, *Introduction to Complex Analysis*, Chelsea, New York, 1982.

- W. F. Osgood, *Functions of a Complex Variable*, G. E. Stechert, New York, 1942.

- G. Polya and G. Latta, *Complex Variables*, John Wiley & Sons, New York, 1974.

- J. Pierpont, *Functions of a Complex Variable*, Ginn & Co., Boston, 1912.

- S. Saks and A. Zygmund, *Analytic Functions*, Nakl. Polskiego Tow. Matematycznego, Warsaw, 1952.

- G. Sansone, *Lectures on the Theory of Functions of a Complex Variable*, P. Noordhoff, Groningen, 1960.

- V. I. Smirnov and N. A. Lebedev, *Functions of a Complex Variable; Constructive Theory*, MIT Press, Cambridge, 1969.

Modern Texts

- A. Beardon, *Complex Analysis: The Argument Principle in Analysis and Topology*, John Wiley & Sons, New York, 1979.

- C. Berenstein and R. Gay, *Complex Variables: An Introduction*, Springer-Verlag, New York, 1991.

- R. P. Boas, *An Invitation to Complex Analysis*, Random House, New York, 1987.

- R. Burckel, *Introduction to Classical Complex Analysis*, Academic Press, New York, 1979.

- J. B. Conway, *Functions of One Complex Variable*, 2nd ed., Springer-Verlag, New York, 1979.

- J. Duncan, *The Elements of Complex Analysis*, John Wiley & Sons, New York, 1969.

- S. D. Fisher, *Complex Variables*, 2nd ed., Brooks/Cole, Pacific Grove, 1990.

- A. R. Forsyth, *Theory of Functions of a Complex Variable*, 3rd ed., Dover, New York, 1965.

- A. O. Gel'fond, *Residues and Their Applications*, Mir Publishers, Moscow, 1971.

- M. Heins, *Complex Function Theory*, Academic Press, New York, 1969.

- E. Hille, *Analytic Function Theory*, 2nd ed., Chelsea, New York, 1973.

- W. Kaplan, *A First Course in Functions of a Complex Variable*, Addison-Wesley, Cambridge, 1953.

- S. Lang, *Complex Analysis*, 3rd ed., Springer-Verlag, New York, 1993.

- N. Levinson and R. M. Redheffer, *Complex Variables*, Holden-Day, San Francisco, 1970.

- J. D. Logan, *Applied Mathematics*, 2nd ed., John Wiley and Sons, New York, 1997.

- A. I. Markushevich, *Theory of Functions of a Complex Variable*, Prentice-Hall, Englewood Cliffs, 1965.

- Jerrold Marsden, *Basic Complex Analysis*, Freeman, San Francisco, 1973.

- G. Mikhailovich, *Geometric Theory of Functions of a Complex Variable*, American Mathematical Society, Providence, 1969.

- R. Narasimhan, *Complex Analysis in One Variable*, Birkhäuser, Boston, 1985.

- T. Needham, *Visual Complex Analysis*, Oxford University Press, New York, 1997.

- J. Noguchi, *Introduction to Complex Analysis*, American Mathematical Society, Providence, 1999.

- B. Palka, *An Introduction to Complex Function Theory*, Springer, New York, 1991.

- R. Remmert, *Theory of Complex Functions*, Springer-Verlag, New York, 1991.

- B. V. Shabat, *Introduction to Complex Analysis*, American Mathematical Society, Providence, 1992.

- M. R. Spiegel, *Schaum's Outline of the Theory and Problems of Complex Variables*, McGraw-Hill, New York, 1964.

Applied Texts

- M. Abramowitz and I. A. Stegun, *Handbook of Mathematical Functions*, Dover, New York, 1965.

- W. Derrick, *Complex Analysis and Applications*, 2nd ed., Wadsworth, Belmont, 1984.

- A. Erdelyi, *The Bateman Manuscript Project*, McGraw-Hill, New York, 1954.

- A. Kyrala, *Applied Functions of a Complex Variable*, John Wiley and Sons, 1972.

- W. R. Le Page, *Complex Variables and the Laplace Transform for Engineers*, McGraw-Hill, New York, 1961.

References

[**AHL**] L. V. Ahlfors, *Complex Analysis*, 3rd ed., McGraw-Hill, New York, 1971.

[**BLK**] B. Blank and S. G. Krantz, *Calculus*, Key Curriculum Press, Emeryville, CA, 2005.

[**BCH**] W. Brown and R. V. Churchill, *Complex Variables and Applications*, 6th ed., McGraw-Hill, New York, 1991.

[**CCP**] Carrier, G., Crook, M., and Pearson, C., *Functions of a Complex Variable: Theory and Technique*, McGraw-Hill, New York, 1966.

[**COH**] R. Courant and D. Hilbert, *Methods of Mathematical Physics*, Wiley-Interscience, New York, 1953.

[**DER**] W. Derrick, *Introductory Complex Analysis and Applications*, Academic Press, New York, 1972.

[**FAK**] H. Farkas and I. Kra, *Riemann Surfaces*, Springer-Verlag, New York, 1971.

[**FIF**] S. D. Fisher and John Franks, The fixed points of an analytic self-mapping, Proc. AMS, 99(1987), 76–79.

[**FOL**] G. B. Folland, *Fourier Analysis and its Applications*, American Mathematical Society, Providence, RI, 2009.

[**GRK**] R. E. Greene and S. G. Krantz, *Function Theory of One Complex Variable*, 2nd ed., American Mathematical Society, Providence, RI, 2002.

[**HEN**] P. Henrici, *Applied and Computational Complex Analysis*, John Wiley & Sons, New York, 1974-1981.

[**HER**] I. Herstein, *Topics in Algebra*, Xerox, Lexington, 1975.

[**HES**] Z. He and O. Schramm, Koebe uniformization and circle packing, *Ann. of Math.* 137(1993), 3612–401.

[**HOR**] L. Hörmander, *Notions of Convexity*, Birkhäuser, Boston, MA, 1994.

[**HUN**] T. Hungerford, *Abstract Algebra: An Introduction*, 2nd ed., Brooks Cole, Pacific Grove, CA, 1996.

[**KAT**] Y. Katznelson, *An Introduction to Harmonic Analysis*, John Wiley & Sons, New York, 1968.

[**KOB**] Kober, H., *Dictionary of Conformal Representations*, 2nd ed., Dover Publications, New York, 1957.

[**KRA1**] S. G. Krantz, *A Panorama of Harmonic Analysis*, Mathematical Association of America, Washington, D.C., 1991.

[**KRA2**] S. G. Krantz, *Real Analysis and Foundations*, 4th ed., CRC Press, Boca Raton, FL, 2016.

[**KRA3**] S. G. Krantz, *Complex Analysis: The Geometric Viewpoint*, 2nd ed., Mathematical Association of America, Washington, D. C., 2004.

[**KRA4**] S. G. Krantz, *Cornerstones of Geometric Function Theory: Explorations in Complex Analysis*, Birkhäuser Publishing, Boston, MA, 2005.

[**LES**] K. Leschinger, Über fixpunkte holomorpher Automorphismen, *Manuscripta Math.*, 25 (1978), 391-396.

[**MAS**] B. Maskit, The conformal group of a plane domain, Amer. J. Math., 90 (1968), 7112–722.

[**NEH**] Nehari, Z., *Conformal Mapping*, Dover Publications, New York, 1952.

[**PAS**] N. Papamichael and N. Stylianopoulos, *Numerical Conformal Mapping: Domains Decomposition and the Mapping of Quadrilaterals*, World Scientific Publishing, Hackensack, NJ, 2010.

[**PRA**] R. Pratrap, *Getting Started with MatLab*, Oxford University Press, Oxford, 1999.

[**RUD1**] W. Rudin, *Principles of Mathematical Analysis*, 3rd ed., McGraw-Hill, New York, 1976.

[**RUD2**] W. Rudin, *Real and Complex Analysis*, McGraw-Hill, New York, 1966.

[**SASN**] E. B. Saff and E. D. Snider, *Fundamentals of Complex Analysis for Mathematics, Science, and Engineering*, 2nd ed., Prentice-Hall, Englewood Cliffs, 1993.

[**SIK**] G. B. Simmons and S. G. Krantz, *Differential Equations: Theory, Technique, and Practice*, McGraw-Hill, New York, 2006.

[**STA**] R. Stanley, *Algebraic Combinatorics: Walks, Trees, Tableaux, and More*, Springer, New York, 2013.

[**STW**] E. M. Stein and G. Weiss, *Introduction to Fourier Analysis on Euclidean Spaces*, Princeton University Press, Princeton, NJ, 1971.

[**THO**] G. B. Thomas, *Thomas's Calculus*, 11th ed., Addison-Wesley, Reading, MA, 2004.

[**ZWI**] Zwillinger, D., *et al.*, *CRC Standard Mathematical Tables and Formulae*, 30th ed., CRC Press, Boca Raton, 1996.

Index

Milton Keynes UK
Ingram Content Group UK Ltd.
UKHW051931141024
449569UK00027B/1443